Engineering Materials

This series provides topical information on innovative, structural and functional materials and composites with applications in optical, electrical, mechanical, civil, aeronautical, medical, bio- and nano-engineering. The individual volumes are complete, comprehensive monographs covering the structure, properties, manufacturing process and applications of these materials. This multidisciplinary series is devoted to professionals, students and all those interested in the latest developments in the Materials Science field, that look for a carefully selected collection of high quality review articles on their respective field of expertise.

Indexed at Compendex (2021) and Scopus (2022)

Arnold C. Alguno · Rey Y. Capangpangan ·
Gerard G. Dumancas · Arnold A. Lubguban ·
Roberto M. Malaluan · Rolen Brian P. Rivera

Gold Nanoparticles

Green Synthesis, Characterization, and Multifaceted Applications

Springer

Arnold C. Alguno
Department of Materials and Resources Engineering and Technology
Mindanao State University-Iligan Institute of Technology
Iligan City, Philippines

Gerard G. Dumancas
Honors College and the Department of Chemistry
North Carolina A&T State University
Greensboro, NC, USA

Roberto M. Malaluan
Department of Chemical Engineering
Mindanao State University-Iligan Institute of Technology
Iligan City, Philippines

Rey Y. Capangpangan
Department of Physical Sciences and Mathematics
Mindanao State University at Naawan
Naawan, Philippines

Arnold A. Lubguban
Department of Chemical Engineering
Mindanao State University-Iligan Institute of Technology
Iligan City, Philippines

Rolen Brian P. Rivera
Mindanao State University-Iligan Institute of Technology
Iligan City, Philippines

ISSN 1612-1317 ISSN 1868-1212 (electronic)
Engineering Materials
ISBN 978-981-96-6564-8 ISBN 978-981-96-6565-5 (eBook)
https://doi.org/10.1007/978-981-96-6565-5

This Springer imprint is published by the registered company Springer Nature Singapore Pte Ltd.
The registered company address is: 152 Beach Road, #21-01/04 Gateway East, Singapore 189721, Singapore

Preface

In the growing list of products developed with the aid of nanotechnology, *Gold Nanoparticles: Green Synthesis, Characterization, and Multifaceted Applications* is among the first to examine in detail the development and application of one of its most important materials: gold nanoparticles (AuNPs). Ranging from the artistic crafts to the life-saving pharmaceuticals, and from environmental protection to electronics and other industries fields, AuNPs are among the most versatile and functional nanoparticles available. These properties are thoroughly examined in this book, which also highlights their size-dependent optical properties, high surface-to-volume ratio, and good biocompatibility, documenting their usefulness across various scientific disciplines.

The principle of sustainable development is central to this work. It focuses on the green synthesis of AuNPs using plant extracts, bacteria, and fungi as reducing and stabilizing agents, thereby presenting an environmentally friendly and scalable approach. Based on the principles of green chemistry, these methods demonstrate the potential of nanotechnology to evolve without compromising the integrity of the environment and society.

This book also provides an overview of important characterization techniques essential developing AuNPs for a wide range of applications, including advanced imaging and photothermal therapy. The application of analytical tools such as UV-Vis spectroscopy, transmission electron microscopy, and X-ray diffraction is discussed to offer supporting evidence and to demonstrate the high degree of accuracy and effectiveness of the research.

Moreover, the book explores the use of AuNPs in advanced applications including biosensors, environmental remediation, nanostructures, hybrid nanostructures, targeted drug delivery, and real-time pollution monitoring. In this context, the book presents a forward-looking perspective on how scientific advancement can align with sustainability principles to achieve meaningful innovation. It envisions the progression of nanotechnology as inherently connected with the pursuit of a sustainable future.

Iligan City, Philippines — Arnold C. Alguno
Naawan, Philippines — Rey Y. Capangpangan
Greensboro, USA — Gerard G. Dumancas
Iligan City, Philippines — Arnold A. Lubguban
Iligan City, Philippines — Roberto M. Malaluan
Iligan City, Philippines — Rolen Brian P. Rivera

Competing Interests The authors have no competing interests to declare that are relevant to the content of this manuscript.

Contents

About the Authors

Arnold C. Alguno is a Professor in the Department of Materials and Resources Engineering and Technology and Director for the Office of Research Management at Mindanao State University-Iligan Institute of Technology (MSU-IIT), Philippines. He conducts research in semiconductor physics, solar cell technology, quantum dots, sensor technology, nanotechnology, and advanced functional materials. Dr. Alguno received his M.Sc. in Physics at MSU-IIT and Ph.D. in Physics in Tohoku University in Japan, where he also completed his postdoctoral fellowship. His research focuses on the synthesis and characterization of nanomaterials and polymeric materials, contributing to technological advancements in material science and nanotechnology.

Rey Y. Capangpangan serves as a Professor of Chemistry at Mindanao State University at Naawan (MSUN), Philippines. He is a nationally recognized researcher and educator specializing in catalysis for environmental applications and the development of advanced nanostructured catalysts. Dr. Capangpangan has led numerous high-impact projects on the synthesis, characterization, and performance evaluation of catalyst materials aimed at environmental remediation, pollution control, and sustainable energy solutions. His pioneering work in environmental catalysis addresses critical challenges in water and air pollution, establishing him as a leading voice in the field of applied environmental catalysis. Dr. Capangpangan's research integrates chemistry, materials science, and engineering, fostering interdisciplinary approaches environmental sustainability and green technology innovation. His commitment to research, teaching, and community engagement underscores his influential role in advancing applied catalytic technologies for environmental sustainability.

Gerard G. Dumancas is an Associate Dean of Research and Program Innovation (Honors College), Professor of Chemistry at North Carolina Agricultural and Technical State University. He was previously an Associate Professor of Chemistry and the Director of the NSF Robert Noyce Teacher Scholarship Program at the University of Scranton in Scranton, PA. He received his B.S. in Chemistry from the University of the Philippines in 2005 and his Ph.D. in Analytical Chemistry (Chemometrics) from

Oklahoma State University in 2012. He completed his postdoctoral training at the Oklahoma Medical Research Foundation in the area of statistical genetics and bioinformatics. His research has been recognized by the American Oil Chemists' Society, the Society for Laboratory Automation and Screening, the NSF Noyce Scholarship Program, the NSF S-STEM Program, and the LSU LIFT2. He was previously named as an NSF Program to Empower Partnerships with Industry Fellow as well as a Visiting Scientist of the Department of Energy's Joint Bioenergy Institute and the US National Renewable Energy Laboratory, and the Berekely Lab in Berkeley, California.

Arnold A. Lubguban is presently appointed as Professor and Senior Researcher at the Mindanao State University-Iligan Institute of Technology (MSU-IIT), in the Philippines. Before joining MSU-IIT in 2015, he served in a short-term Balik-Scientist engagement with the Department of Science and Technology (DOST) of the Republic of the Philippines in 2014, delivering lectures and training on bio-based polymer processing to academic institutions and local industries. He received his degree in Chemical Engineering in 1999 from the University of the Philippines-Los Baños (UPLB) and obtained both his M.S. and Ph.D. in Chemical Engineering from the University of Missouri in 2005 and 2009, respectively. He pursued postdoctoral research in industrial polymer processing in 2009 at the Ohio State University, focusing on the conversion of lignocellulosic biomass into value-added polyols and polyurethanes. He has spearheaded multiple research initiatives funded by USAID-STRIDE and DOST-PCIEERD on the development and processing of sustainable, bio-based polymers for advanced industrial applications.

Roberto M. Malaluan originated from San Jose, Batangas, Philippines, and pursued his undergraduate education at Mindanao State University-Marawi, receiving a B.S. degree in Chemical Engineering 1983. He worked as an engineering instructor at Mindanao State University-Iligan Institute of Technology (MSU-IIT) for several years before attending Tohoku University in Sendai, Japan, where he received his Ph.D. under the guidance of Professor Kunio Arai in 1995. He became a full Professor of Chemical Engineering at MSU-IIT in 2004. He has held multiple leadership roles on supercritical carbon dioxide extraction of selected natural products and indigenous species for drug discovery and industrial applications under the Department of Agriculture (DA) and the Department of Science and Technology (DOST) of the Republic of the Philippines. He currently serves as a Senior Faculty Member, engaged in teaching and conducting research on sustainable polymers.

Rolen Brian P. Rivera earned his Doctor of Philosophy degree in Physics at Mindanao State University-Iligan Institute of Technology (MSU-IIT), Philippines. He is a Postdoctoral Research Fellow on the CHED-LAKAS-funded Colorimetric Project at MSU-IIT. His work focuses on the development of a colorimetric sensor for detecting food spoilage. He holds both Bachelor of Science and Master of Science degrees in Physics from MSU-IIT, where he gained extensive expertise in the synthesis and physicochemical characterization of metal nanoparticles for colorimetric applications. His research advances the design of low-cost, rapid-response nanosensors to enhance food safety, spoilage detection, and quality assurance monitoring systems.

Chapter 1
Introduction

Abstract Gold nanoparticles (AuNPs) are newly explored nanoscale materials that possess numerous advantages over the conventional materials due to their novel optical, chemical, and biological properties. Widely applied in biosensing, imaging, and targeted drug delivery because of their localized surface plasmon resonance (LSPR) property, they hold significant value in medicine and diagnostics. The shape, and stability of nanoparticles can be precisely controlled by various synthesis methods, including chemical, physical, and green synthesis, to optimize their functional performance. The use of plant extracts and microorganisms in green synthesis eliminates the need for toxic substances, thereby reducing waste and enhancing the biocompatibility of the resulting nanoparticles. However, challenges remain, including limited scalability, lack of reproducibility, and environmental safety concerns. Recent advancements in microfluidics, artificial intelligence-driven optimization, and hybrid nanomaterials are addressing these issues, unlocking their potential for industrial applications. This chapter highlights the role of AuNPs in advancing sustainable nanotechnology, precision medicine, and environmental remediation. It also discusses the development of AuNPs and various synthesis methods, with particular emphasis on green synthesis and its critical contribution to the progression of sustainable nanotechnology.

1.1 Overview

Gold nanoparticles (AuNPs) play crucial role in nanotechnology due to their unique optical, chemical, and biological properties. These nanoscale materials possess a high surface area and are already utilized in medicine, electronics, and environmental science applications. AuNPs exhibit distinct optical absorption or scattering phenomenon known as localized surface plasmon resonance (LSPR), which can be effectively applied in diagnostics, imaging, and controlled drug delivery. The stability, functionality, and biocompatibility of AuNPs depend on the method of synthesis, whether chemical, physical or green. Inspired by Richard Feynman's

A. C. Alguno et al., *Gold Nanoparticles*, Engineering Materials,
https://doi.org/10.1007/978-981-96-6565-5_1

vision of nanotechnology articulated in 1959, the field has advanced considerably, with AuNPs now considered essential in modern scientific applications. Progress in scanning probe microscopy and controlled synthesis has opened new opportunities for their integration across diverse disciplines. Nevertheless, challenges persist in terms of standardization, scalability for industrial production, and environmental implications. As nanotechnology continues to evolve, responsible innovation and adherence to ethical considerations remain critical. This chapter discusses the significance of AuNPs, their synthesis, and their future potential in shaping emerging technologies while fostering mutual advancement of science and environment sustainability.

1.2 Scope and Significance of Gold Nanoparticles (AuNPs)

Gold nanoparticles (AuNPs) are small nanoscale fragments of gold that have fascinated scientists for many years, dating back to ancient craftsmen who created the extraordinary gold ruby glass. Also referred to as nanoparticles, AuNPs range in size from 1 and 100 nanometers and exhibit properties markedly different from those of their bulk counterparts due to their small size and high surface area to volume ratio. Their synthesis via chemical, biological, or green chemistry approaches enables precise control over their size, shape, and functionality to unprecendented degree. Since the conceptual inception of nanotechnology in 1959, the development of AuNPs has progressed from decorative uses to critical roles in modern science. This transformation is part of the broader nanotechnology movement focused on designing and manipulating materials at the sub-100 nanometer scale. The long-term stability of AuNPs in air and their chemical inertness have facilitated their application across numerous fields, including chemistry, biology, and material science [1]. AuNPs possess tunable optical characteristics governed by the LSPR effect, which depends on the nanoparticles' size, shape, concentration, and surrounding environment.

On the nanoscale level, gold displays a remarkable richness of color that ranges from red to black to sapphire, driven by structural variations at the atomic level. This versatility extends to a wide array of morphologies such as nanorods, nanostars, nanocubes, and nanospheres, each engineered for specific applications [2]. Due of their excellent biocompatibility and tuneable optical properties, AuNPs are transforming biomedicine and a pivotal role in the advancement of drug delivery systems, diagnostics, imaging, and therapeutic strategies [3]. Beyond healthcare, AuNPs exhibit exceptional adaptability in applications across agriculture, electronics, textiles, and environmental sensing. Their chemophysical properties can be deliberately harnessed to expand their utility, positioning AuNPs as indispensable materials in the modern landscape of scientific and technological innovation.

1.2.1 Overview of Nanotechnology and Its Evolution

Nanotechnology, which involves the manipulation of matter at the nanoscale, has become a transformative force in nearly all areas of science and technology. Its conceptual origin can be traced to Richard Feynman's 1959 lecture entitled "*There's Plenty of Room at the Bottom,*" in which he proposed the possibility of manipulating individual atoms to create new materials. The 1980s were a period of rapid advancement, during which instruments such as the Scanning Tunneling Microscope (STM) and Atomic Force Microscope (AFM) were developed. These tools were made possible to visualize and interact with matter at the atomic level. At the nanoscale, the materials exhibit properties that differ significantly from their bulk counterpart due to size-dependent effect. For example, AuNPs display characteristics distinct from those of larger gold particles. While initial uses of AuNPs were primarily decorative, they are now in medicine, electronics, and environmental science. Various synthesis methods, including chemical reduction, laser ablation, and green biosynthesis, allow the precise control over size, shape, and surface chemistry of AuNPs for targeted applications. The foundational contribution was made by Michael Faraday in 1857 when he prepared colloidal gold and observed its unique optical properties, which subsequently inspired extensive scientific inquiry.

Today, nanotechnology plays a critical role in biomedicine, catalysis, and environmental monitory [4]. The significance of AuNPs lies in their biocompatibility, optical tunability, and ease of functionalization, which have enabled advancements in cancer therapy, drug delivery systems, and biosensing technologies. The inherent interdisciplinary nature of nanoscience foster collaboration among chemists, physicists, biologists, and materials scientists to develop novel technologies with the potential to significantly improve human life. Nanotechnology has garnered increasing global attention, as evidenced by the exponential growth in related publications over the last two decades. The trend indicates that the number of research papers containing the term "nano" is rising at a near- linear rate and may exceed 30,000 by 2025, as shown in Fig. 1.1. This review presents foundational concepts in nanotechnology and examines the properties and applications of transition metal nanoparticles.

Despite the rapid progress, significant challenges remain in understanding the health and environmental effects of emerging nanoscale materials. The field also faces obstacles in commercialization, including the need for standardization, improving production technologies, and comprehensive toxicity assessment. As governments address these concerns, it is essential to proceed cautiously in the continued development of nanotechnology. In this rapidly expanding domain, responsible allocation of research funding and rigorous evaluation of potential health and environmental risks are critical. Maintaining a balance between innovation and ethics, as well as between risk and safety, is vital to realizing the full potential of nanotechnology and its benefits for society.

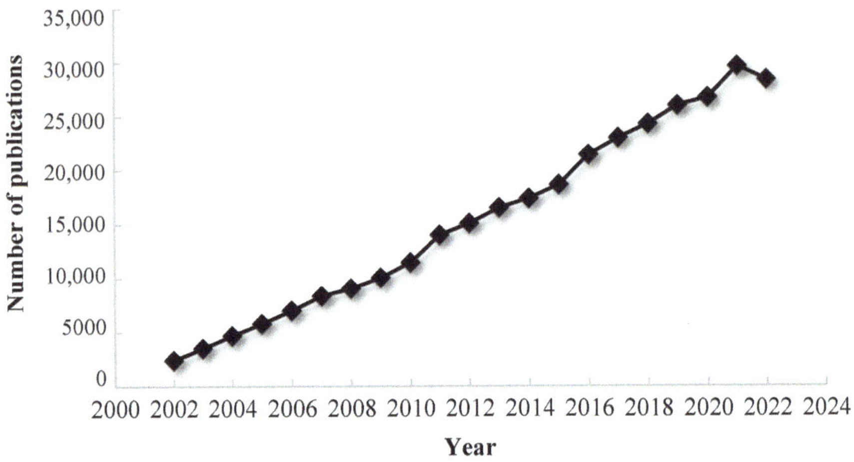

Fig. 1.1 The number of publications in the field of nanotechnology based on the Web of Science data (as of 20 May 2023). Adapted from [5], Fig. 1.1, https://doi.org/10.3390/molecules28134932. Licensed under CC BY 4.0 (https://creativecommons.org/licenses/by/4.0/)

1.2.2 Importance of AuNPs in Nanoscience Due to Their Unique Optical, Chemical, and Biological Properties

AuNPs have become significant components in the field of nanoscience due to their unique optical properties. At the core of their functionality is the LSPR, which results in strong and tunable optical absorption through excitation of conduction electrons [6, 7]. The LSPR effect is illustrated in Fig. 1.2 to show how incident light induces surface plasmon, also referred to as surface plasmon resonance, on nanoparticles whose dimensions are comparable to the wavelength of the incoming light. The resonance frequency depends on factors such as electron density, effective mass, and charge distribution, and it can be modulated by altering the material composition and nanostructure morphology to enhance the electric field in the near-surface region.

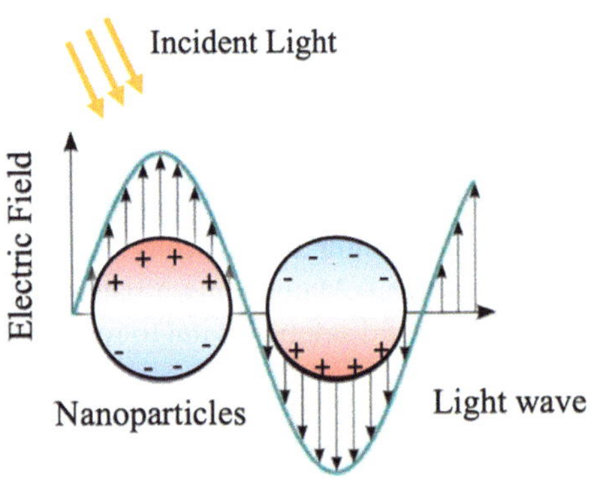

Fig. 1.2 An illustrative representation of localized surface plasmon resonance (LSPR) in metallic nanoparticles, highlighting the oscillation of free electrons on their surfaces. Adapted from [16], Fig. 1.2, https://doi.org/10.3390/mi14071393. Licensed under CC BY 4.0 (https://creativecommons.org/licenses/by/4.0/)

The size and shape of AuNPs are critical in determining their optical resonance. For instance, 20 nm diameter spheres exhibit LSPR at around 520 nm, whereas nanorods exhibit a red shift exceeding 700 nm [8]. These optical properties are essential for colorimetric sensing applications, as the pronounced color change from red (dispersed state) to blue (aggregated state) serves a clear indicator of analyte presence. This shift is based on interparticle plasmon coupling, enabling extremely high sensitivity with molar extinction coefficients on the order of 10^9 M^{-1} cm^{-1}. As a result, AuNPs can be employed for the detection of biomolecules and pollutants at nanomolar concentrations [9].

The synergy between optical characteristics and practical applications is exemplified by surface-enhanced Raman scattering (SERS), which utilizes the amplified electromagnetic field generated by LSPR to enhance Raman signal by six orders of magnitude or more. This enables the detection of substances such as pesticides and heavy metals [10, 22]. For instance, AuNPs have demonstrated sufficient sensitivity to detect SARS-CoV-2 RNA at femtomolar level, which is highly significant for diagnostic assays and environmental monitoring [8, 22].

Their optical advantages, AuNPs are chemically stable and reactive. Their selectivity can be enhanced by conjugation with ligands such as DNA, antibodies and enzymes, making them highly suitable for targeted drug delivery and gene therapy [10]. For example, PEGylated AuNPs can remain in circulation for 48 h, significantly extending their half-life and enabling more efficient delivery with reduced toxicity [11].

Due to their ability to shift oxidation state from Au^0 to Au^+ to Au^{3+}, AuNPs can mimic enzymatic catalytic reactions, achieving turnover numbers greater than 100 s^{-1} [9]. Their biocompatibility further enhances their relevance in precision medicine, with applications in treating cancer, cardiovascular diseases, and arthritis, improving therapeutic outcomes with minimal side effects [10]. Moreover, AuNPs are employed in environmental science for detecting heavy metals and degrading organic pollutants through energetically favorable catalytic processes, supporting the advancement of green chemistry. The development of green synthesis methods using plant extracts and microorganisms has significantly improved the sustainability of AuNP production by eliminating toxic reagents and enhancing the environmental compatibility of the process. This approach extends the applicability of AuNPs across various sectors, including medicine, food technology, and diagnostics [8].

1.3 Current Trends and Challenges in AuNP Research

Traditional methods in synthesizing AuNPs typically involve the chemical or physical reduction of gold salts in a dispersive medium, often in water. However, these methods pose significant challenges for industrial-scale production. They require dangerous, toxic, and costly reagents, making them unsuitable for large-scale applications.

Furthermore, they generate substantial volumes of chemical waste, which is ecologically detrimental. The ongoing search for sustainable alternatives has been a driving force behind the increasing focus on ecological responsibility. Researchers have thus developed green synthesis approaches, also referred to as biogenic synthesis, phytofabrication, or biosynthesis, that avoid the use of toxic chemicals and promote clean, cost-effective production. These methods allow the precise control over nanoparticle size and morphology, which is essential for optimizing their performance in applications such as medicine, electronics, and catalysis [12, 13]. In addition, these green strategies offer a potential reduction in production cost. At the global level, governments are implementing stricter regulations on hazardous substance, thereby increasing the demand for sustainable methods of AuNP production. Investments are being directed toward cleaner and safer technologies to ensure both economic viability and environmental compliance.

1.3.1 Existing Synthesis Methods: Chemical, Physical, and Biological Approaches

AuNPs can be synthesized through chemical, physical and biological methods, each of which has its advantages and limitations. These synthesis techniques influence scalability, reproducibility, environmental impact, and the physicochemical properties of the resulting nanoparticles.

Chemical Methods. Chemical synthesis is the most common approach for producing AuNPs. This method involves low-energy reducing agents such as borohydrides, phosphites, alcohols, and hydrogen peroxide, which reduces gold ions (Au^{3+}) to elemental gold (Au^0) [14].

Using citrate as the reducing agent, the Turkevich method allows control over nanoparticles size by adjusting parameters such as pH, temperature, and reactant concentrations. Another example is the Brust-Schiffrin method, which employs organic solvents to produce monodisperse AuNPs suitable for biomedical applications. However, chemical synthesis presents significant challenges. Environmental and health concerns. Additionally, residual contaminants may effects the surface plasmon resonance of AuNPs, thereby compromising their performance in sensing applications. Although chemical methods are efficient and reproducible, their environmental impact necessitates the development of more sustainable alternative.

Physical Methods. Physical approaches to AuNP synthesis include laser ablation, sputtering, and thermal decomposition, although they remain limited in application [17]. Among these, laser ablation is particularly notable. In this method, high-energy lasers vaporize bulk gold in a liquid medium, eliminating the need for chemical reducing agents. This process yields highly pure nanoparticles free from chemical contaminations. Nevertheless, the high energy requirements and reliance on complex equipment pose limitations. Furthermore, these techniques often offer limited control over nanoparticle size and morphology. As a result, physical methods are generally less suitable for large-scale production compared to chemical synthesis.

Biological Methods. Biological synthesis provides a green alternative by utilizing natural reducing and stabilizing agents such as plant extracts, fungi, and bacteria to reduce gold ions [18]. Plant derived materials, though their intrinsic phytochemicals, act simultaneously as reduction and stabilizing agents, eliminating the need for toxic chemicals. This approach is cost-effective, biocompatible, and well suited for biomedical applications. However, biological synthesis faces challenges in terms of reproducibility and precise control over nanoparticle properties, which limits scalability and industrial application. On going research seeks to refine these techniques to improve consistency and commercial viability.

Each synthesis method offers specific benefits that make it more appropriate for particular applications. While chemical synthesis remains the most widely used. Increasing emphasis on sustainability is shifting attention towards biological approachess. An ideal synthesis method should combine efficiency, reproducibility, and environmental sustainability. The field is progressively advancing toward cleaner and more efficient production techniques through the development of green chemistry and nanobiomaterial-based strategies. With continued research and innovation, AuNP synthesis may transision to more sustainable practices, aligning technological advancement with ecological responsibility.

1.3.2 Environmental Concerns and the Shift Towards Green Synthesis

The adverse effects of the conventional AuNP synthesis have been investigated in recent studies. The traditional methods utilize hazardous reducing agents such as sodium borohydride, which are harmful to the environment and result in the generation of toxic waste [19, 20]. Although physical methods produce less chemical waste, they are energy-intensive and inefficient in terms of carbon emissions. These disadvantages highlight the necessity for developing eco-friendly and sustainable alternatives.

Green synthesis has emerged as a promising approach. This method employs natural biological systems such as plant extracts, bacteria, and fungi to stabilize and shape nanoparticles under relatively mild temperature conditions. Biomolecules such as proteins, polyphenols, and flavonoids act as natural reducing and stabilizing agents. Among the green approach, plant extract-based synthesis is particularly attractive due to its simplicity and environmental compatibility. In contrast to chemical methods, which are typically conducted under harsh conditions and often require external stabilizers, green synthesis offers a more sustainable alternative.

However, the application of green synthesis still faces several challenges. Controlling the formation of nanoparticles with the desired size and shape remains a issue. The physicochemical properties of AuNPs depend significantly on the type of plants used, its growth conditions, and the method of extracting the biological agents, all of which can vary considerably. Additionally, while green synthesis is effective at the laboratory scale, its scalability for industrial production remains limited. Efforts are ongoing to improve reproducibility, efficiency, and scalability of these methods.

The concept of green synthesis aligns well with the principles of sustainable development and meets industry demand for environmentally friendly technologies. Researchers are working on establishing standardized biological synthesis protocols and optimizing production conditions for large- scale applications. This work is essential to ensure the quality, stability, and safety of biosynthesized AuNPs, using advanced characterization techniques such as high-resolution electron microscopy and spectroscopy [21].

Recent studies demonstrate that biosynthesized AuNPs can outperformed conventionally synthesized nanoparticles. For instance, the use of *Brassicaceae* plant extracts has yielded nanoparticles of various shapes and sizes, including triangular, spherical, pentagonal, hexagonal, and cuboidal morphologies, ranging from 10 to 200 nm. These findings indicate that green synthesis enables better control over nanoparticle size and shape, while also enhancing the functional properties of the AuNPS.

In conclusion, green synthesis represents a significant advancement in nanoparticle production. It is not only cost-effective and environmentally friendly but also enables the generation of highly pure nanoparticles. The current limitations may be overcome through the development of improved synthesis protocols and scalable production systems. The future of AuNP synthesis lies in the integration of sustainability and innovation to produce nanomaterials that are both safe and effective.

1.4 Objectives of the Book

The book contains seven chapters, each of which explores important areas related to AuNPs. It begins with an introductory chapter discusses AuNP synthesis, with particular emphasis on green synthesis approaches, which are emerging as key topics of interest in contemporary scientific literature, showcasing these sustainable techniques. This chapter aims to generate interest within the scientific community and highlight the vast potential of AuNPs in diverse applications. It explains the significance of green synthesis in the context of AuNPs development using accessible theories and terminologies. The objective is to integrate these foundational elements of knowledge to present a cohesive perspective on AuNP applications across multiple disciplines.

Subsequent chapters aim to keep researchers informed the current development in the field of AuNPs. Chapters two and three expand on the fundamental principles introduced earlier, defining the working mechanisms of nanoparticles with specific focus on AuNPs. Chapters four and five explores the broad spectrum of AuNPs applications, including in biomedicine, environmental science, and electronics, thereby underscoring the versatility and relevance. Open-minded readers with comprehensive understanding on the diverse industrial uses of AuNPs.

The penultimate and final chapters build upon the previous discussions by exploring future trends and anticipated advances in AuNP applications. These chapters highlight the continued evolution of AuNPs within innovative and emerging fields. In addition to offering a current and original overview of significant nanoparticles and nanocapsules, original contributions from leading experts actively engaged in AuNP research. This ensures that readers are presented with the latest methodologies and pioneering applications. Collectively, the book as an educational resource and a reference for future research initiatives in the field of nanotechnology.

1.4.1 To Provide an In-Depth Understanding of AuNP Synthesis, Properties, and Applications

Consequently, even though the body of literature on metal nanoparticles has expanded significantly, the number of related manuscripts continues to grow, driven by the development of these materials into a central focus within nanotechnology. This book intends to provide comprehensive overview of AuNPs in terms of synthesis, properties, and wide-ranging applications. It is suitable for self-study, with every chapter beginning with clearly written preparatory material to guide the reader through the concepts to be covered. The organization of topics is intentionally straight forward to enhance readability, while the broader goal is to introduce readers to the science of AuNPs and equip them with the foundational knowledge needed to engage with this rapidly evolving interdisciplinary field.

One of the central themes of the book is the diversity of chemically synthesis methods available for AuNPs. These include classical approaches such as the Turkevich method and the Brust-Schiffrin technique, as well as physical and emerging continous production methods.

The book also emphasizes distinctive properties of AuNPs, particularly their unique optical property resulting from LSPR. These properties are critical for applications in biosensing, drug delivery and imaging. A significant portion of the book is devoted to medical and biological applications, with a particular focus on targeted drug delivery systems. AuNPs can be functionalized to enhance biocompatibility, making them promising candidates for therapeutic applications such as photothermal cancer theraphy and gene delivery.

While the book discusses a range of applications, its objective is to present a holistic view of the current state and future potential of AuNPs. It emphasizes their role in driving innovation across various scientific domains by serving as a bridge between fundamental research and real-world applications.

1.4.2 To Explore the Role of Green Synthesis Techniques

Environmental sustainability has emerged as a growing paradigm in nanoparticle synthesis, underscoring the urgent need for green processes. Conventional approaches often involve significant energy consumption and the use of toxic hydro-organic solvents, leading to environmental concerns and elevated production costs. The extreme conditions typically associated with these methods, such as high pressure and high temperature, can also promote nanoparticle aggregation, potentially causing complications during the dispersion. Therefore, there is an critical need for sustainable and eco-friendly production routes for colloidal nanoparticles that are both effective and environmentally responsible.

In this context, green synthesis has emerged as a promising alternative. These methods emphasize the use of environmentally benign solvents and safe reaction conditions with low energy requirements. The use of natural reducing agents such as plant extracts, bacteria, and fungi facilitae the synthesis of AuNPs without undesirable side effects often encountered in classical synthesis. Plant extracts are particularly attractive, as the phytochemicals they contain can simultaneously act as reducing agent (for metal salt reduction) and stabilizing agent (to prevent reaggregation). This dual funationality enables a straightforward synthesis process without the need for additional chemical reagents. Within the scope of this work, the rapid synthesis of well-dispersed AuNPs was achieved under ambient conditions by simply mixing gold salts with plant extracts.

Despite these advantages, challenges remain, particularly regarding variability in particle size and shape, which stems from the differences in composition of the biological agents used in green synthesis. Efforts to standardize these processes to improve reproducibility are ongoing. Current scientific developments in green synthesis are advancing toward the establishment of effective methodologies for the chemical, microscopic, and spectroscopic characterization of nanoparticles with potential pharmacological effects. In conclusion, green synthesis represents a more environmentally friendly alternative to conventional methods, offering pathway to produce high-performance nanoparticles with reduce toxicity and enhanced sustainability.

1.4.3 To Highlight Emerging Applications and Future Directions

Research on AuNPs have made substantial progress, with their applications now spanning a wide range of fields. This expansion suggests a promising future with numerous anticipated developments. This section highlights some of the most actively developing areas for AuNPs, including biomedicine, electronic materials, and environmental remediation.

In medicine, AuNPs serve as a valuable tools for the advancement of diagnostics and therapeutics. Their versatility of nanocarriers enables the development of targeted drug delivery systems that enhances therapeutic efficacy while reduced toxicity. Furthermore, AuNPs function effectively as imaging agents, allowing for real-time monitoring of therapeutic responses. A notable application is photothermal theraphy, in which AuNPs efficiently convert light into heat to destroy cancer cells, underscoring their significant therapeutic potential.

Another innovative area of development is the use of AuNPs in electronic materials. Researchers are incorporating these nanoparticles into electronic and photonic devices both organic and inorganic substrates. The unique properties of AuNPs enhance the performance and functionality of the devices, leading to advancements in electronics technologies.

AuNP research also holds great promise for environmental remediation. These nanoparticles are being engineered to remove various pollutants, including heavy metals and organic pollutants, thereby contributing to cleaner ecosystems. Their catalytic properties are particularly promising for clean water purification, addressing critical issues related to environmental sustainability.

Looking ahead, AuNPs are expected to play a pivotal role in addressing major societal challenges, including improving healthcare accessibility and promoting sustainable practices. There is a growing demand for interdisciplinary collaboration among experts in biology, chemistry, materials science, and medicine to fully realize the potential of AuNPs. As the field ventures beyond traditional boundaries, careful attention to ethical considerations and governance will be essential to ensure the responsible development of AuNP-based technologies. Through the synergistic integration of scientific disciplines, AuNPs are poised to achieve their full potential as plasmonic materials, offering innovative solutions to technological challenges and contributing meaningfully to societal problems.

References

1. Hammami, I., Alabdallah, N.M.: Gold nanoparticles: synthesis properties and applications. J. King Saud Univ. Sci. (2021). https://doi.org/10.1016/j.jksus.2021.101560
2. Sarfraz, N., Khan, I.: Plasmonic gold nanoparticles (AuNPs): properties, synthesis and their advanced energy, environmental and biomedical applications. Chem. Asian J. (2021). https://doi.org/10.1002/asia.202001202
3. Nejati, K., Dadashpour, M., Gharibi, T., Mellatyar, H., Akbarzadeh, A.: Biomedical applications of functionalized gold nanoparticles: a review. J. Cluster Sci., 1–16 (2021). https://doi.org/10.1007/s10876-020-01955-9
4. Malik, S., Muhammad, K., Waheed, Y.: Nanotechnology: a revolution in modern industry. Molecules (2023). https://doi.org/10.3390/molecules28020661
5. Szczyglewska, P., Feliczak-Guzik, A., Nowak, I.: Nanotechnology-general aspects: a chemical reduction approach to the synthesis of nanoparticles. Molecules **28**(13), 4932 (2023). https://doi.org/10.3390/molecules28134932
6. Khurana, K., Jaggi, N.: Localized surface plasmonic properties of Au and Ag nanoparticles for sensors: a review. Plasmonics (2021). https://doi.org/10.1007/s11468-021-01381-1

7. Zhao, J., Xue, S., Ji, R., Li, B., Li, J.: Localized surface plasmon resonance for enhanced electrocatalysis. Chem. Soc. Rev. (2021). https://doi.org/10.1039/D1CS00237F
8. Sadiq, Z., Safiabadi Tali, S.H., Hajimiri, H., Al-Kassawneh, M., Jahanshahi-Anbuhi, S.: Gold nanoparticles-based colorimetric assays for environmental monitoring and food safety evaluation. Crit. Rev. Anal. Chem. **54**(7), 2209–2244 (2024). https://doi.org/10.1080/10408347.2022.2162331
9. Ma, W., Zhan, Y., Zhang, Y., Mao, C., Xie, X., Lin, Y.: The biological applications of DNA nanomaterials: current challenges and future directions. Signal Transduct. Target. Ther. **6**(1), 351 (2021). https://doi.org/10.1080/10408347.2022.2162331
10. Mitchell, S., Qin, R., Zheng, N., Pérez-Ramírez, J.: Nanoscale engineering of catalytic materials for sustainable technologies. Nature Nanotechnol. (2021). https://doi.org/10.1038/s41565-020-00799-8
11. Ding, Y., Wang, Y., Hu, Q.: Recent advances in overcoming barriers to cell-based delivery systems for cancer immunotherapy. Exploration (2022). https://doi.org/10.1002/EXP.20210106
12. Patil, T., Gambhir, R., Vibhute, A., Tiwari, A.P.: Gold nanoparticles: synthesis methods, functionalization and biological applications. J. Cluster Sci. (2023).https://doi.org/10.1007/s10876-022-02287-6
13. Qiao, J., Qi, L.: Recent progress in plant-gold nanoparticles fabrication methods and bio-applications. Talanta (2021). https://doi.org/10.1016/j.talanta.2020.121396
14. Hassan, H., Sharma, P., Hasan, M.R., Singh, S., Thakur, D., Narang, J.: Gold nanomaterials–the golden approach from synthesis to applications. Mater. Sci. Energy Technol. **5**, 375–390 (2022). https://doi.org/10.1016/j.mset.2022.09.004
15. Ahmed, S.F., Mofijur, M., Rafa, N., Chowdhury, A.T., Chowdhury, S., Nahrin, M.,... Ong, H.C.: Green approaches in synthesising nanomaterials for environmental nanobioremediation: technological advancements, applications, benefits and challenges. Environ. Res. **204**, 111967 (2022). https://doi.org/10.1016/j.envres.2021.111967
16. Wu, J.Z., Ghopry, S.A., Liu, B., Shultz, A.: Metallic and non-metallic plasmonic nanostructures for LSPR sensors. Micromachines **14**(7), 1393 (2023). https://doi.org/10.3390/mi14071393
17. Ali, F., Hamza, M., Iqbal, M., Basha, B., Alwadai, N., Nazir, A.: State-of-art of silver and gold nanoparticles synthesis routes, characterization and applications: a review. Z. Phys. Chem. **236**(3), 291–326 (2022). https://doi.org/10.1515/zpch-2021-3084
18. Rónavári, A., Igaz, N., Adamecz, D.I., Szerencsés, B., Molnar, C., Kónya, Z.,... Kiricsi, M.: Green silver and gold nanoparticles: Biological synthesis approaches and potentials for biomedical applications. Molecules **26**(4), 844 (2021). https://doi.org/10.3390/molecules26040844
19. Baytar, O., Şahin, Ö, Ekinci, A.: Effect of environmentally friendly and efficient metal-free hydrochars as catalysts on sodium borohydride hydrolysis. Fuel (2023). https://doi.org/10.1016/j.fuel.2023.128308
20. Wang, T., Jiang, T., Zhang, H., Zhao, Y.: Advances in catalysts for hydrogen production by methanolysis of sodium borohydride. Int. J. Hydrogen Energy **47**(32), 14589–14610 (2022). https://doi.org/10.1016/j.ijhydene.2022.02.173
21. Sargazi, S., Laraib, U., Er, S., Rahdar, A., Hassanisaadi, M., Zafar, M.N.,... Bilal, M.: Application of green gold nanoparticles in cancer therapy and diagnosis. Nanomaterials **12**(7), 1102 (2022). https://doi.org/10.3390/nano12071102
22. Myalowenkosi, S., Sebastien, B., Mikhael, B., Jean-Marc, J., Krishna, B.: A Review of Gold and Silver Nanoparticle-Based Colorimetric Sensing Assays Advanced Engineering Materials **19**(12) 1700270 (2017). https://doi.org/10.1002/adem.v19.12

Chapter 2
Synthesis of Gold Nanoparticles

Abstract Gold nanoparticles (AuNPs) are currently among the most well-known nanomaterials, with numerous applications in medicine, catalysis, and environmental sciences, and other fields. Their synthesis has been achieved through physical, chemical, and biological methods, each differing in mechanism and requirements. Physical preparation techniques such as laser ablation and evaporation–condensation can produce high-purity AuNPs, but they require expensive and sophisticated equipment. The Turkevich–Frens and Brust–Schiffrin methods are examples of chemical synthesis techniques. These methods offer precise control over nanoparticle size and stability, which is particularly important in biomedical applications. Green synthesis, which utilizes plant extracts and microorganisms, provides a sustainable alternative that generates minimal or no toxic waste, making it suitable for biological systems. The size, shape, and stability of AuNPs are influenced by the synthesis parameters such as pH, temperature, and reagent concentration. However, challenges remain, including limited scalability, poor reproducibility, and potential environmental impact. Promising innovations to address these issues include microfluidics, artificial intelligence driven optimization, and the development of hybrid nanomaterials. This chapter will also highlight the various sources of AuNPs and their synthesis methods, with a focus on sustainable approaches and their implications for the future of nanotechnology and industry.

2.1 Overview

Gold nanoparticles (AuNPs) are at the forefront of nanoscale science, possessing a unique set of remarkable optical, electronic and catalytic properties that make them useful in a wide range of fields, including medicine and environmental diagnostics. Several physical, chemical, and biological methods are employed to synthesize these particles at the nanoscale, with each approach offering distinct advantages and versatility. Physical techniques such as laser ablation and sputtering enable precise control over the size and shape of AuNPs through thermal energy, which is critical for advanced optoelectronic applications [1, 2]. One specific example is the use of

A. C. Alguno et al., *Gold Nanoparticles*, Engineering Materials,
https://doi.org/10.1007/978-981-96-6565-5_2

sodium citrate, an inexpensive one-pot chemical route, and sodium borohydride in the Brust-Schiffrin method, which is effective in the iodine reduction of perrhenate and remains widely used in biomedical applications [3].

Additionally, biological synthesis using plant extracts and microorganisms with green chemistry principles, addressing environmental concerns while reducing hazardous by-products [4, 5]. By finely tuning synthesis conditions such as pH, temperature, and the choice of reductant, researchers can tailor the size, shape, and surface chemistry of AuNPs to meet the demands of various applications, including targeted drug delivery, biosensing, and environmental remediation. This adaptability makes AuNPs indispensable in advancing sustainable technologies and revolutionizing scientific and industrial applications, establishing them as a fundamental element in the progression of contemporary science [6, 51]. AuNPs have been synthesized through a variety of approaches (Fig. 2.1), each with specific advantages and applications. Physical methods, including laser ablation, are noted for their precision and efficiency, while chemical techniques encompass a broad range of reactions that yield nanoparticles with desireable properties. Biological substances used in green synthesis offer environmentally friendly alternatives to conventional techniques, harnesing the vast potential of natural nanoparticle production. Therefore, understanding the underlying mechanisms these synthesis strategies can facilitate the exploration of new application areas across diverse fields such as medicine, electronics, and environmental science.

2.2 Overview of Synthesis Techniques

AuNPs are highly versatile nanomaterials that can be synthesized through various mechanisms, which are primarily classified into physical and chemical approaches. Physical methods such as evaporation and laser ablation utilize physical processes to generate AuNPs from bulk gold [8, 9]. In evaporation, gold atoms are vaporized and subsequently assemble into nanoparticles depending on the duration of the process. In contrast, laser ablation involves immersing a solid gold substrate in liquid medium, where a laser is used to ablate the gold surface, producing gold atoms that nucleate to form nanoparticles [10]. Although physical methods offer high purity and tunable nanoparticle characteristics, they typically require sophisticated and expensive instruments.

By comparison, chemical method are widely used for AuNP synthesis due to their ability to closely control nanoparticle properties. A key chemical approach allows the production of AuNPs across wide size range by adjusting the ratio of gold salts and reducing agent. Another commonly used chemical technique is polyol synthesis, which efficiently reduces gold ions using suitable reducing agents [11]. Recently, eco-friendly synthesis techniques based on biological entities, including plant extracts, have gained popularity. These green synthesis methods supports cleaner, non-toxic, and sustainable nanoparticle production. Understanding the different synthesis routes

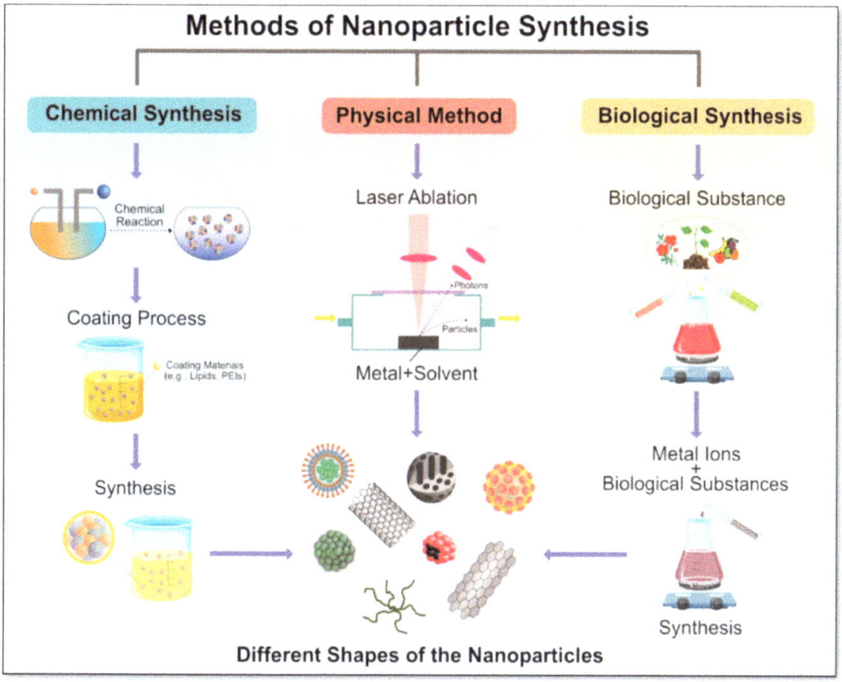

Fig. 2.1 Considering the various AuNP synthesis techniques to highlight the specific advantages of chemical, physical, and biological approaches. Adapted from [7], Fig. 2.1, https://doi.org/10.3390/catal11121494. Licensed under CC BY 4.0 (https://creativecommons.org/licenses/by/4.0/)

is essential, as they directly influence the optical properties and functional applications of AuNPs. These methods continue to evolve to improve efficiency and expand their utility in areas such as drug delivery and advanced bioimaging [12, 13].

2.2.1 *Physical Methods (Laser Ablation, Evaporation–Condensation)*

Laser ablation is an established method for the synthesis of nanoparticles, based on a top-down approach that involves bombarding high-energy laser pulses onto a gold substrate submerged in an aqueous medium [14]. This interaction generates a plasma plume that facilitates both the nucleation and growth of the nanoparticles. Laser ablation is considered a highly pure technique for producing nanoparticle dispersions, as it does not require chemical agents, making it particularly promising for biomedical and electronics applications. Parameters such as laser wavelength, pulse duration, and energy density are critical variables that can be precisely controlled to adjust particle size and distribution. However, the resource-intensive method requires

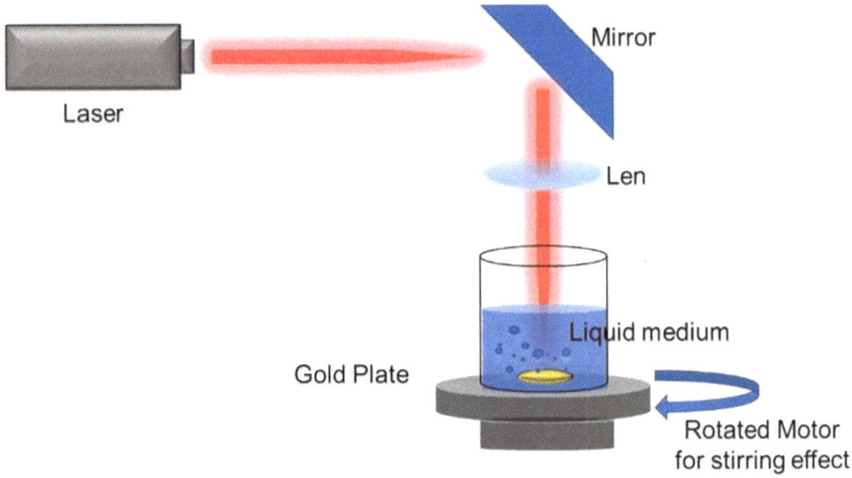

Fig. 2.2 Laser Ablation method in synthesizing AuNPs. Adapted from [10], Fig. 2.2, https://doi.org/10.3390/ma15030875. Licensed under CC BY 4.0 (https://creativecommons.org/licenses/by/4.0/)

complex equipment and significant energy output, which may limit its widespread adoption in industrial settings. The laser ablation process shown in Fig. 2.2, adapted from the graphical abstract published by [10].

In contrast, the evaporation–condensation method synthesizes nanoparticles by volatilizing gold inside an inert gas at high temperatures, followed by cooling to induce condensation [15, 16]. This technique is recognized for its ability to produce homogeneous nanoparticles with well-defined structural and optical properties, makes it suitable for specific applications. However, it shares similar limitations in terms of high energy consumption and equipment requirements. Both methods represent examples of physical synthesis approaches used to produce high-purity nanoparticles, which are essential in applications where material purity is critical [17]. Nevertheless, the need for more advanced technology systems that improve energy efficiency and cost-effectiveness remains a significant challenge scale up these techniques for industrial application. The field of nanoparticle synthesis continues to evolve, and there is potential for the emergence of novel method involving electron and ion beam technologies.

2.2.2 Chemical Methods (Turkevich-Frens, Brust-Schiffrin, Martin Methods)

Nanoscience and technology continues to investigate AuNPs for applications in diagnostics, drug delivery, and catalysis. Key chemical methods contributing to the

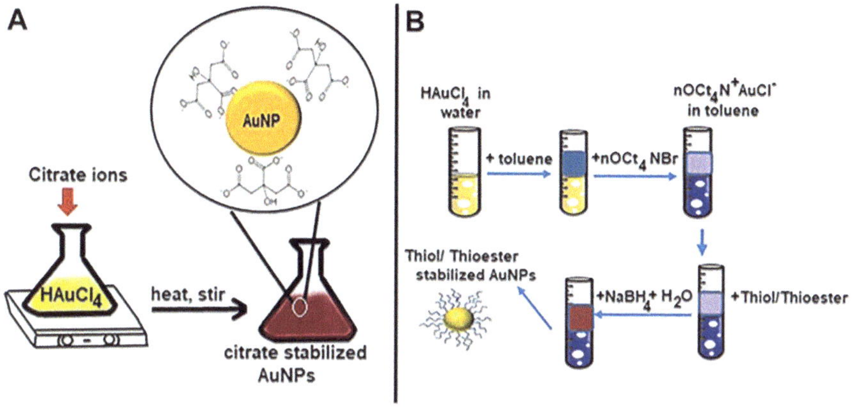

Fig. 2.3 (A) Schematic representation of the Turkevich method for synthesizing AuNPs and (B) sequential steps involved in the Brust–Schiffrin method for the synthesis of thiol-stabilized AuNPs. Adapted from [55], Fig. 2.3, https://doi.org/10.2147/IJN.S279094. Licensed under CC BY 4.0 (https://creativecommons.org/licenses/by/4.0/)

synthesis of nanoparticles with controlled sizes and properties include the Turkevich-Frens, Brust-Schiffrin, and Martin methods. The Turkevich-Frens method, shown in Fig. 2.3, is a classical chemical approach that employs citrate ions to reduce chloroauric acid ($HAuCl_4$), resulting in the formation of stable spherical AuNPs [2]. This method allows for the selective adjustment of particle size (12–15 nm) by varying temperature, pH, and the citrate – gold ratio, which is particularly important for bioconjugates that require particle sizes greater than 30 nm to achieve optimal biological activity.

On the other hand, the Brust-Schiffrin method, shown in Fig. 2.3b, synthesizes thiol-capped AuNPs in organic solvents using phase transfer catalyst. This method produces nanoparticles with improved stability and narrower size distribution, which enhances their suitability for functionalization [18]. In contrast, the Martin method involves controlled decompartmentalization using mild reducing agents, ascorbic acid, to tailor the properties of AuNPs for application in biosensing and catalysis [19]. Evaluating the environmental sustainability of AuNP synthesis processes and applying the principles of green chemistry, particularly the reduction in the use of toxic and hazardous substances, can enhance the relevance of AuNPs while minimizing their environment impact.

2.3 Biological Synthesis Approaches

Several technologies are available for preparing metal nanoparticles, among which biological methods have recently gained popularity as environmentally friendly alternatives. These are the use of plants, bacteria, fungi, and natural extracts for the biological synthesis of AuNPs offers an eco-friendly and cost-effective process that can be carried out at room temperature [20, 21]. This rapid, one-pot synthesis utilizes phytochemicals and enzymes present in plant extracts to reduce metal ions and stabilize the resulting nanoparticles. Green synthesis provides advantages such as high efficiency, minimal purification requirements, and enhanced biocompatibility [22]. Overall, biological synthesis supports current trends in technology and medicine that emphasize the sustainable production of nanoparticles.

2.3.1 *Plant-Mediated Synthesis*

Forest of plants is a natural resources of signifcant ecological and economic importance. They serve as sources of both reducing and stabilizing agents in this nanoparticle synthesis, making plant-mediation synthesis a novel method and environmentally friendly approach for producing AuNPs [23, 24]. Researchers extract essential phytochemicals, such as flavonoids, alkaloids, terpenoids and phenolics, from various parts of plants, including leaves, roots, bark and flowers. A stepwise procedure is followed to prepare plants materials for AuNPs synthesis. Since the synthesis process requires a clean environment, the plant parts must be thoroughly washed to remove any contaminants. After cleaning, the plant materials are gently dried and ground into a fine powder to serve as the initial feedstock for sequential extraction. The desired phytochemicals are then isolated through aqueous extraction, followed by filtration to remove larger particulates, resulting in a refined extract with a characteristic muddy residue. Finally, the plant extract is mixed with the metal salt ($HAuCl_4$), initiating the formation of AuNPs. This process produces plant-derived AuNPs with a variety of morphologies as shown in Fig. 2.4.

This series of steps involves washin, drying, boiling, and extracting phytochemicals from plant materials, which chemically reduces gold ions (Au^{3+}) too metallic gold (Au^0). This approach is praised for its adaptability and energy efficiency, and often proceeding under ambient conditions without the need of complex procedures. The formation of nanoparticles that are both scalable and cost-effective for industrial applications is indicated by a visible color change in the reaction mixture, typically shifting from light yellow to a ruby red or purple.

Additionally, this method aligns well with the principle of green chemistry due to the absence of toxic reagents [25, 26]. The resulting AuNPs are biocompatible, enhancing their suitability for biomedical applications, such as drug delivery, imaging, and photothermal therapy. Several plants have demonstrated the ability to produce AuNPs with consistent characteristics, functioning as non-toxic

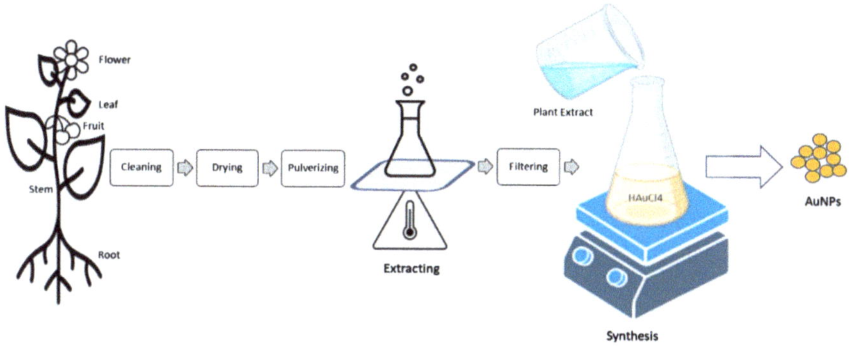

Fig. 2.4 Phytochemical isolation, filtering, and subsequent interaction with $HAuCl_4$ results in the formation of AuNPs with diverse morphologies during the aqueous extraction and synthesis of plant-based AuNPs. Adapted from [27], Fig. 2.2, https://doi.org/10.3897/pharmacia.71.e112322. Licensed under CC BY 4.0 (https://creativecommons.org/licenses/by/4.0/)

reducing agents in this synthesis process. For example, AuNPs have been synthesized both intracellularly and extracellularly using *Medicago sativa* as the reducing agent. Similary, *Mangifera indica* leaf extract has been shown to effectively mediate the reduction of $H[AuCl_4]^{3+}$. Various metals have also been to produce nanoparticles of different sizes using *Lagerstroemia* (4.9 nm–100 nm), *Antiaris toxicaria* (20 nm) and *Syzygium cumini* (100 nm).

2.3.2 *Bacterial and Fungal Synthesis*

The use of plants and mushrooms in the environment-friendly synthesis of AuNPs represents a potentially significant advancement in nanotechnology. The metabolites secreted by these living organisms can facilitate the bioreduction of gold ions into nanoparticles, offering a promising green alternative to conventional synthesis methods [28, 22]. This approach marks a groundbreaking step in nanotechnology by utilizing low-cost materials to enable continuous production. It leverages biological components that not only simplify the synthesis process but also promote biomass growth. As a result, it supports the generation of size-uniform particles while maintaining scalability. Among various biological methods, the use of microbes is one of the most effective strategies for nanoparticles synthesis. This innovative approach employs a range of biological entities, including bacteria, fungi, and algae, as illustrated in Fig. 2.5.

Bacteria, in particular, facilitate the extracellular reduction of gold salts through enzymes such as nitrate reductase. Well-studied and diverse bacterial species, including *Pseudomonas aeruginosa* and *Bacillus subtilis,* are known to biosynthesize AuNPs with tunable properties, shapes and sizes, making them useful in medicine, catalysis and sensing applications. Unlike bacteria, fungi represent a

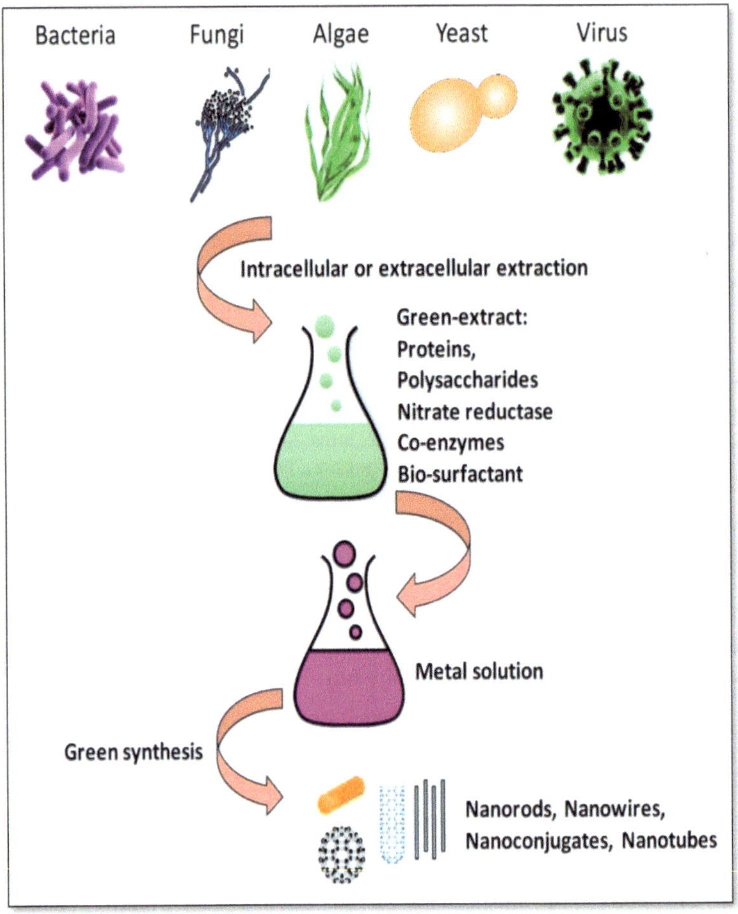

Fig. 2.5 By using microorganisms such as bacteria, fungus, and algae, the biological synthesis of nanoparticles offers a scalable and cost-effective manufacturing method. Adapted from [7], Fig. 2.4, https://doi.org/10.3390/catal11121494. Licensed under CC BY 4.0 (https://creativecommons.org/licenses/by/4.0/)

more capable heterotropic system with the potential to excrete higher quantities of extra-cellular enzymes, thereby enhancing the stabilization and bioreduction of nanoparticles.

Scientific fungal species, such as *Aspergillus niger* and *Fusarium oxysporum*, have been widely studied in nanoparticle research [28, 22]. As the synthetic process progresses, the reaction can be monitored by changes in the color of the reaction medium, which is typically analyzed using UV–Vis spectroscopy. Although both bacterial and fungal approaches are relatively simple and environmentally benign, challenges remain. These includes prolong synthesis times and potential pathogenicity of biological agents, neccessitating careful strain selection and process optimization. This is further improved by using organic reducing agents derived directly from the microorganisms, thereby reducing potential adverse effects [22, 29]. Advancement in this field, along with continued exploration of microbial capabilities expected to open new avenues for complex bioprocesses and contribute to the sustainable production of a wide variety of metal nanoparticle systems.

2.4 Green Synthesis of AuNPs

Over the past years, the biogenic synthesis of AuNPs from Au (III) precursors has emerged as one of the most promising ecofriendly approaches, utilizing reducing and stabilizing agents derived from various plans extracts, organisms, and microorganisms. This method minimizes environmental pollution and waste production, aligning with the principles of green chemistry [30, 31]. Sustainable practices are advanced through green synthesis by producing nanoscale materials without toxic by-products. This approach is widely applied across multiple fields, particularly in sensors, drug delivery, and bioimaging, highlighting the importance of ecological preservation and pollution reduction through the green synthesis of nanoparticles [30].

2.4.1 Definition and Importance of Green Synthesis

The reduction of chloroauric salt using the indigenous resources such as microorganisms and plant extract to produce AuNPs represents a novel and eco-friendly method for synthesizing AuNPs through green synthesis [32, 33]. This green approach significantly reduces ecological footprints while yielding functional materials. Natural reducing agents, including polyphenols and proteins, can be employed to produce biocompatible nanoparticles for medical applications such as drug delivery, cancer treatment, and diagnostics. Moreover, this method substantially decreases the generation of toxic by-products and waste, aligning with global sustainability goals and the growing demand for environmentally safe materials [31, 34]. As nanobiotechnology continues to evolve, green synthesis represents itself as a sustainable route for meeting industrial requirements while adhering to environmental principles and practices.

2.4.2 *Advantages of Green Methods Over Traditional Approaches*

The advantages of the green synthesis of AuNPs, compared to conventional physical and chemical methods, including a significant reduction in the use of toxic reagents and lower energy consumption, resulting in enhanced safety and sustainability [32, 33]. This technique generates non-harmful by-products from natural compounds, contributing to a lower carbon footprint [35–37]. Moreover, the simple and cost-effective use of plant extracts enables reactions to proceed at room temperature, producing biocompatible nanoparticles. As ongoing research continues to refine green synthesis protocols, their potential for scaling up to commercial applications is increasingly recognized. However, scalability remains a challenge despite the growing demand for eco-friendly methods relevant to bioimaging and diagnostic applications.

A comparison between green synthesis methods and conventional approach for AuNPs synthesis is presented in Table 2.1, which provides comparative analysis of their advantages, disadvantages, cost, and applications. Green synthesis promotes environmental sustainability by utilizing natural compounds found in plants and microbes to produce biocompatible nanoparticles, yielding non toxic by-products suitable for medical applications such as bioimaging. Nonetheless, challenges such as limited scalability and slower reaction rates persist. In comparison, the conventional synthetic methods offer precise control and scalable production but often rely on hazardous precursors and intermediates. Therefore the development of green alternatives, particularly those based on bio-inspired nanoparticle synthesis, is imperative to meet both industrial and environmental objectives.

2.5 Factors Influencing AuNP Synthesis

The performance of AuNPs in these novel applications is highly dependent on the size, morphology and functional properties, all of which are governed by a number of crucial synthetic parameters. These physicochemical characteristics are strongly influenced by factors such as pH, temperature, and reagent concentration. For example, pH variations can alter the charge of citrate ions, which act as stabilizers. This, in turn, may affect their ability to prevent nanoparticle aggregation, often reflected in a distinct color change from red to blue, characteristic of plasmonic coupling [1, 5].

Temperature also plays a significant role ; higher temperatures generally enhances nucleation and growth rates, thereby affecting particle uniformity and morphology. Additionally, the molar ratio of reagents must be carefully controlled, as it determines the size distribution of AuNPs. Precise stoichiometric control is essential when

Table 2.1 Comparative analysis of green synthesis and conventional methods for synthesizing gold nanoparticles, highlighting their respective advantages, disadvantages, cost, and applications.

Aspect	Green Synthesis Methods	Traditional Approaches
Advantages	• Eco-safe, employing natural substances (like plant extracts and microbes) [32, 33] • Non-toxic residue minimizes environmental impact [36, 37] • Reactions can be conducted under ambient conditions with low energy consumption • Biocompatible nanoparticles for medical uses such as bioimaging and diagnosis	• Highly scalable, as many protocols are well established [6] • Provides a better control over particle size, shape, and morphology • Fast reaction rates with reproducible results
Disadvantages	• Scalability is limited because it relies on biological materials [33] • Lack of uniformity and reproducibility in particle size and morphology vs. conventional methods • Less-reactive, which may affect industrial production	• Use of hazardous reagents, labor-intensive and risking health and environment [2] • High energy requirements such as heating and vacuum systems • Creates toxic byproducts that need specialized disposal [38]
Cost	• Economical from availability of raw biological resources [35]	• Increased costs associated with expensive reagents and specialized equipment [39]
Applications	• Great for biodegradable and biocompatible applications in tissue engineering, drug delivery, etc	• Used frequently for catalysis, electronics and industrial-scale nanotechnology [6]

synthesizing for specific applications [15, 40]. For this reason, the use of strict experimental controls, combined with a comprehensive understanding of these influential factors, is vital for optimizing synthesis methodologies.

Through systematic investigation aimed at the rapid identification of AuNPs with optimal characteristics, such as biocompatibility or tunable optical properties, researchers can unlock new potential applications in cutting-edge fields, including bioimaging, diagnostics, and catalysis. This area of research is particularly important, as the science of AuNP synthesis becomes increasingly complex, opening new avenues for the advancement of nanotechnology in both industrial and biomedical domains [6, 50].

2.5.1 Impact of pH, Temperature, and Concentration on Nanoparticle Size and Shape

The synthesis of AuNPs is heavily influenced by three main parameters, pH, temperature, and concentration, which directly affect the nucleation and growth processes and ultimately determine the final size and shape of the nanoparticles. pH is a critical factor that can significantly alter the surface charge and the kinetics of the reduction process. Generally, under acidic conditions, slower nucleation can lead to the formation of larger and less uniform particles. In contrast, alkaline conditions accelerate nucleation, resulting in smaller and more uniformly shaped nanoparticles, highlighting the importance of pH modulation. This pH-dependent behavior is particularly important in green synthesis methods that utilize plant-derived reducing agents, which are highly sensitive to pH changes [41].

Temperature is another crucial factor in nanoparticle formation. Elevated temperatures increase the kinetics of reduction and nucleation energy, promoting the formation of smaller and more homogeneous particles. Conversely, lower temperatures tend to favor particle agglomeration [42]. In addition, the concentration of the gold precursor must be carefully considered. According to reference [43], higher precursor concentrations can lead to nucleation and the formation of larger particles, whereas lower concentrations promote controlled growth and yield nanoparticles with more desireable morphologies and dimensions.

These parameters should be carefully optimized for the targeted synthesis of AuNPs with properties suited to specific applications, including biosensing, drug delivery, diagnostics, and environmental monitoring. By deliberately adjusting synthesis conditions, it is possible to tailor the physicochemical charateristics of AuNPs to meet the needs of various advanced technological applications [51, 52]. Understanding and controlling these factors is essential in enhancing the efficiency, precision, and scalability of nanomaterial synthesis in increasingly complex domains.

2.6 Challenges and Future Trends in AuNP Synthesis

In the face of these very challenging barriers, the synthesis and applications of AuNPs are likely to witness a breakthrough in evolution in the near future. The major challenge lies in the variable reproducibility of the synthesis of AuNPs synthesis, which limits the production of uniformly dispersed, monodisperse nanoparticles, an essential requirement for precision-based applications such as early diagnostics and high-demand industrial sectors [44, 45]. Size and morphology inconsistencies remain a critical obstacle to achieving commercial scalability and standardization.

Additionally, the environmental impact resulting from nanomaterial production has become a growing concern, as conventional synthesis techniques frequently generates toxic by-products that pose risks to human health and environmental balance, thereby complicating waste management efforts [50–52]. In response, the concept of green synthesis techniques has emerged, utilizing biological pathways (plant- or microbe-based) to minimize toxic waste and exploit renewable resources [53]. Looking ahead, it is encouraging to observe the integration of advanced computational models, AI- driven optimizations, and workflow management systems that are enhancing synthesis protocols at multiple levels, including scalability and reproducibility [6]. The characterization of nanoparticles is also become more precise through real-time monitoring and in situ analysis, providing researchers with unprecedented control over nanoparticle properties.

In the long term, AuNPs are poised to revolutionize various industry sectors through their incorporation into multifunctional devices, enabling a wide range of applications in biosensors, therapeutics, and environmental monitoring. Given the growing demands in nanotechnology, it essential for researchers to adopt scalable, cost-effective, and environmentally responsible strategies, allowing AuNPs to serve as a sustainable solution with global impact [6, 54].

2.6.1 Addressing Scalability, Reproducibility, and Environmental Impact

Scalability constitutes a fundamental challenge in the industrialization of nanoparticles, requiring that the synthesis method be capable of consistently generating large quantities. Colloidal methods provide high-quality nanoparticles [46], but demanded substantial power input and specialized equipment. Conventional methods have known limitations, whereas green synthesis methods utilize renewable biological resources and may offer scalable alternative. Nevertheless, achieving consistent high yields, these methods remains difficult and requires comprehensive knowledge of the underlying mechanisms of nanoparticle formation. In addition, reproducibility is a critical factor to investigate, as biological reducing agents can vary and lead to nanoparticles with different properties [47, 48].

To address this issue, it is necessary to establish standardized extraction protocols and implement automated monitoring systems to ensure reproducibility across production batches. Environmental concerns also plays a significant role; conventional synthesis commonly involves toxic precursors and generates hazardous waste, while green synthesis is generally less harmful to the environment, despite certain limitations [49]. Current research emphasize the promise of hybrid strategies that combine traditional and green approaches, along with the integration of artificial intelligence for real-time process optimization, as a viable solutions to the challenges of scalability and reproducibility. By combining interdisciplinary expertise and deepening our understanding of nanoparticle across varying environments, researchers can develop innovative production methods that promote the sustainable application of gold nanoparticles in diverse fields.

References

1. Hammami, I., Alabdallah, N.M.: Gold nanoparticles: synthesis properties and applications. J. King Saud Univ. Sci. (2021).https://doi.org/10.1016/j.jksus.2021.101560

2. Quinson, J., Aalling-Frederiksen, O., Dacayan, W. L., Bjerregaard, J.D., Jensen, K.D., Jørgensen, M.R.,... Jensen, K.M.: Surfactant-free colloidal syntheses of gold-based nanomaterials in alkaline water and mono-alcohol mixtures. Chem. Mater. **35**(5), 2173–2190 (2023).https://doi.org/10.1021/acs.chemmater.3c00090
3. Qi, K., Zhuang, C., Zhang, M., Gholami, P., Khataee, A.: Sonochemical synthesis of photocatalysts and their applications. J. Mater. Sci. Technol. **123**, 243–256 (2022). https://doi.org/10.1016/j.jmst.2022.02.019
4. Ielo, I., Rando, G., Giacobello, F., Sfameni, S., Castellano, A., Galletta, M.,... Plutino, M.R.: Synthesis, chemical–physical characterization, and biomedical applications of functional gold nanoparticles: a review. Molecules **26**(19), 5823 (2021).https://doi.org/10.3390/molecules26195823
5. Sarfraz, N., Khan, I.: Plasmonic gold nanoparticles (AuNPs): properties, synthesis and their advanced energy, environmental and biomedical applications. Chem. Asian J. (2021). https://doi.org/10.1002/asia.202001202.
6. Feldheim, D.L., Foss, C.A.: Metal nanoparticles: synthesis, characterization, and applications. Marcel Dekker (2002).https://doi.org/10.1021/ja015381a
7. Roy, A., Elzaki, A., Tirth, V., Kajoak, S., Osman, H., Algahtani, A., Islam, S., Faizo, N.L., Khandaker, M.U., Islam, M.N., Emran, T.B., Bilal, M.: Biological synthesis of nanocatalysts and their applications. Catalysts **11**(12), 1494 (2021). https://doi.org/10.3390/catal11121494
8. Krishnia, L., Thakur, P., Thakur, A.: Synthesis of nanoparticles by physical route. In: Synthesis and Applications of Nanoparticles, pp. 45–59. Springer Nature Singapore, Singapore (2022). https://doi.org/10.1007/978-981-16-6819-7_3
9. Liang, S.X., Zhang, L.C., Reichenberger, S., Barcikowski, S.: Design and perspective of amorphous metal nanoparticles from laser synthesis and processing. Phys. Chem. Chem. Phys. **23**(19), 11121–11154 (2021). https://doi.org/10.1039/d1cp00701g
10. Mat Isa, S.Z., Zainon, R., Tamal, M.: State of the art in gold nanoparticle synthesisation via pulsed laser ablation in liquid and its characterisation for molecular imaging: a review. Materials (2022). https://doi.org/10.3390/ma15030875
11. Villaverde-Cantizano, G., Laurenti, M., Rubio-Retama, J., Contreras-Cáceres, R.: Reducing Agents in Colloidal Nanoparticle Synthesis–an Introduction (2021). https://doi.org/10.1039/9781839163623-00001
12. Ahmed, S.F., Mofijur, M., Rafa, N., Chowdhury, A.T., Chowdhury, S., Nahrin, M.,... Ong, H.C.: Green approaches in synthesising nanomaterials for environmental nanobioremediation: technological advancements, applications, benefits and challenges. Environ. Res. **204**, 111967 (2022). https://doi.org/10.1016/j.envres.2021.111967
13. Javaid, M., Haleem, A., Singh, R.P., Suman, R., Gonzalez, E.S.: Understanding the adoption of Industry 4.0 technologies in improving environmental sustainability. Sustain. Oper. Comput. **3**, 203–217 (2022). https://doi.org/10.1016/j.susoc.2022.01.008
14. Mukherjee, D.: New Frontiers for heterostructured nanocomposites with interfacial functionalities synthesized via laser ablation synthesis in solution (LASiS). Emerging Mater. Environ. (2023). https://doi.org/10.1007/978-3-031-39470-6_5
15. Hachem, K., Ansari, M.J., Saleh, R.O., Kzar, H.H., Al-Gazally, M.E., Altimari, U.S.,... Kianfar, E.: Methods of chemical synthesis in the synthesis of nanomaterial and nanoparticles by the chemical deposition method: a review. BioNanoScience **12**(3), 1032–1057 (2022).https://doi.org/10.1007/s12668-022-00996-w
16. Malekzadeh, M., Swihart, M.T.: Vapor-phase production of nanomaterials. Chem. Soc. Rev. (2021).https://doi.org/10.1039/D0CS01212B
17. Pustovalov, V.K.: Laser melting, evaporation, and fragmentation of nanoparticles: experiments, modeling, and applications. Nanotechnol. Precision Eng. (2025). https://doi.org/10.1063/10.0034539
18. Singh, H., Bamrah, A., Bhardwaj, S.K., Deep, A., Khatri, M., Brown, R.J.,... & Kim, K.H.: Recent advances in the application of noble metal nanoparticles in colorimetric sensors for lead ions. Environ. Sci. Nano **8**(4), 863-889 (2021).https://doi.org/10.1039/D0EN00963F

19. Guselnikova, O., Lim, H., Kim, H.J., Kim, S.H., Gorbunova, A., Eguchi, M.,... Yamauchi, Y.: New trends in nanoarchitectured SERS substrates: nanospaces, 2D materials, and organic heterostructures. Small **18**(25), 2107182 (2022). https://doi.org/10.1002/smll.202107182
20. Adeyemi, J.O., Oriola, A.O., Onwudiwe, D.C., Oyedeji, A.O.: Plant extracts mediated metal-based nanoparticles: synthesis and biological applications. Biomolecules (2022).https://doi.org/10.3390/biom12050627
21. Soni, V., Raizada, P., Singh, P., Cuong, H.N., Rangabhashiyam, S., Saini, A.,... Nguyen, V.H.: Sustainable and green trends in using plant extracts for the synthesis of biogenic metal nanoparticles toward environmental and pharmaceutical advances: a review. Environ. Res. **202**, 111622 (2021). https://doi.org/10.1016/j.envres.2021.111622
22. Xu, F., Li, Y., Zhao, X., Liu, G., Pang, B., Liao, N.,... Shi, J.: Diversity of fungus-mediated synthesis of gold nanoparticles: properties, mechanisms, challenges, and solving methods. Critical Rev. Biotechnol. **44**(5) (2024), 924-940.https://doi.org/10.1080/07388551.2023.2225131
23. Dheyab, M.A., Oladzadabbasabadi, N., Aziz, A.A., Khaniabadi, P.M., Al-ouqaili, M.T., Jameel, M.S.,... Ghasemlou, M.: Recent advances of plant-mediated metal nanoparticles: synthesis, properties, and emerging applications for wastewater treatment. J. Environ. Chem. Eng., 112345 (2024).https://doi.org/10.1016/j.jece.2024.112345
24. Naaz, R., Siddiqi, W.A., Mohsin, M.: Plant-mediated synthesis of nanomaterials for biosensing. In: Nanoscale Sensors and their Applications in Biomedical Imaging, pp. 177–188. Springer Nature Singapore, Singapore (2024). https://doi.org/10.1007/978-981-97-3144-2_11.
25. Pattoo, T.A.: Flora to nano: sustainable synthesis of nanoparticles via plant-mediated green chemistry. Plant Sci. Arch. (2023). https://doi.org/10.51470/PSA.2023.8.1.12
26. Zaman, W., Ali, S., Akhtar, M.S.: Harnessing the power of plants: innovative approaches to pollution prevention and mitigation. Sustainability (2024).https://doi.org/10.3390/su16231 0587
27. Suliasih, B.A., Budi, S., Katas, H.: Synthesis and application of gold nanoparticles as antioxidants. Pharmacia **71**, 1–19 (2024). https://doi.org/10.3897/pharmacia.71.e112322
28. Qamar, S.U.R., Ahmad, J.N.: Nanoparticles: mechanism of biosynthesis using plant extracts, bacteria, fungi, and their applications. J. Molecular Liquids (2021). https://doi.org/10.1016/j.molliq.2021.116040
29. Verma, J., Kumar, C., Sharma, M., Saxena, S.: Biotechnological advances in microbial synthesis of gold nanoparticles: Optim. Appl., 3 Biotech (2024). https://doi.org/10.1007/s13205-024-04110-7
30. Samuel, M.S., Ravikumar, M., John J.A., Selvarajan, E., Patel, H., Chander, P.S.,... Chandrasekar, N.: A review on green synthesis of nanoparticles and their diverse biomedical and environmental applications. Catalysts **12**(5), 459 (2022).https://doi.org/10.3390/catal12050459
31. Soltys, L., Olkhovyy, O., Tatarchuk, T., Naushad, M.: Green synthesis of metal and metal oxide nanoparticles: principles of green chemistry and raw materials. Magnetochemistry (2021).https://doi.org/10.3390/magnetochemistry7110145
32. Rónavári, A., Igaz, N., Adamecz, D.I., Szerencsés, B., Molnar, C., Kónya, Z.,... Kiricsi, M.: Green silver and gold nanoparticles: biological synthesis approaches and potentials for biomedical applications. Molecules **26**(4), 844 (2021).https://doi.org/10.3390/molecules26040844
33. Santhosh, P.B., Genova, J., Chamati, H.: Green synthesis of gold nanoparticles: an eco-friendly approach. Chemistry (2022).https://doi.org/10.3390/chemistry4020026
34. Aswathi, V.P., Meera, S., Maria, C.A., Nidhin, M.: Green synthesis of nanoparticles from biodegradable waste extracts and their applications: a critical review. Nanotechnol. Environ. Eng. **8**(2), 377–397 (2023). https://doi.org/10.1007/s41204-022-00276-8
35. Dikshit, P.K., Kumar, J., Das, A.K., Sadhu, S., Sharma, S., Singh, S.,... Kim, B.S.: Green synthesis of metallic nanoparticles: applications and limitations. Catalysts **11**(8), 902 (2021). https://doi.org/10.3390/catal11080902

36. Kumar, J.A., Krithiga, T., Manigandan, S., Sathish, S., Renita, A.A., Prakash, P.,... Crispin, S.: A focus to green synthesis of metal/metal based oxide nanoparticles: various mechanisms and applications towards ecological approach. J. Cleaner Produ. **324**, 129198 (2021). https://doi.org/10.1016/j.jclepro.2021.129198
37. Vijayaram, S., Razafindralambo, H., Sun, Y.Z., Vasantharaj, S., Ghafarifarsani, H., Hoseinifar, S.H., Raeeszadeh, M.: Applications of green synthesized metal nanoparticles—a review. Biol. Trace Elem. Res. **202**(1), 360–386 (2024). https://doi.org/10.1007/s12011-023-03645-9
38. Jain, K., Takuli, A., Gupta, T.K., Gupta, D.: Rethinking nanoparticle synthesis: a sustainable approach vs. traditional methods. Chem. Asian J. **19**(21), e202400701 (2024). https://doi.org/10.1002/asia.202400701
39. Adiga, U., Rai, T., Desy, T.M., Adiga, S.: Role of gold nanoparticles as novel agents for cancer diagnosis and therapy. In: Sundram, S., Malviya, R., Koteswara Rao, G.S.N. (Eds.), Metals in Medicine: Volume 2: Metallic Nanoparticles for Biomedical Applications, pp. 123–145. Elsevier (2025). https://doi.org/10.1201/9781003538455.
40. Hadadian, Y., Masoomi, H., Dinari, A., Ryu, C., Hwang, S., Kim, S.,... Yoon, J.: From low to high saturation magnetization in magnetite nanoparticles: the crucial role of the molar ratios between the chemicals. ACS Omega **7**(18), 15996–16012 (2022). https://doi.org/10.1021/acsomega.2c01136
41. Pourali, P., Benada, O., Pátek, M., Neuhöferová, E., Dzmitruk, V., Benson, V.: Response of biological gold nanoparticles to different pH values: Is it possible to prepare both negatively and positively charged nanoparticles? Appl. Sci. **11**(23), 11559 (2021). https://doi.org/10.3390/app112311559
42. Khelfa, A., Nelayah, J., Amara, H., Wang, G., Ricolleau, C., Alloyeau, D.: Quantitative in situ visualization of thermal effects on the formation of gold nanocrystals in solution. Adv. Mater. **33**(38), 2102514 (2021). https://doi.org/10.1002/adma.202102514
43. Wu, K.J., Edmund, C.M., Shang, C., Guo, Z.: Nucleation and growth in solution synthesis of nanostructures–from fundamentals to advanced applications. Progress Mater. Sci. (2022).https://doi.org/10.1016/j.pmatsci.2021.100821
44. Ahmad, A.: Safety and toxicity implications of multifunctional drug delivery nanocarriers on reproductive systems in vitro and in vivo. Front. Toxicol. (2022). https://doi.org/10.3389/ftox.2022.895667
45. Li, X., Luo, R., Liang, X., Wu, Q., Gong, C.: Recent advances in enhancing reactive oxygen species based chemodynamic therapy. Chinese Chem. Lett. (2022).https://doi.org/10.1016/j.cclet.2021.11.048
46. Operti, M.C., Bernhardt, A., Grimm, S., Engel, A., Figdor, C.G., Tagit, O.: PLGA-based nanomedicines manufacturing: technologies overview and challenges in industrial scale-up. Int. J. Pharm. **605**, 120807 (2021). https://doi.org/10.1016/j.ijpharm.2021.120807
47. Dawadi, S., Katuwal, S., Gupta, A., Lamichhane, U., Thapa, R., Jaisi, S.,... Parajuli, N.: Current research on silver nanoparticles: synthesis, characterization, and applications. J. Nanomater. (1), 6687290 (2021).https://doi.org/10.3390/min13010091
48. Ulusoy, U.: A review of particle shape effects on material properties for various engineering applications: from macro to nanoscale. Minerals (2023).https://doi.org/10.3390/min13010091
49. Ahmaruzzaman, M., Mishra, S.R., Gadore, V., Yadav, G., Roy, S., Bhattacharjee, B.,... Kumari, K.: Phenolic compounds in water: from toxicity and source to sustainable solutions-an integrated review of removal methods, advanced technologies, cost analysis, and future prospects. J. Environ. Chem. Eng., 112964 (2024).https://doi.org/10.1016/j.jece.2024.112964
50. Singh, M., & Bhuvaneshwari, M.: (Eds.). Green biosynthesis of nanoparticles: Mechanisms and applications. CABI (2013). https://doi.org/10.1079/9781780642239.0000
51. Mahendra, R.: Yadav Plants as potential synthesiser of precious metal nanoparticles: progress and prospects IET Nanobiotechnology. **7**(3) 117–124 (2013). https://doi.org/10.1049/iet-nbt.2012.0031
52. Pasquale, B., Maria, A., Malvindi, S., Sayaji, K., Rosalia, B., Maria, S., Stefania, S.: Dissolution test for risk assessment of nanoparticles: a pilot study Nanoscale. **9**(19) 6315–6326 (2017). https://doi.org/10.1039/C6NR08131B

53. Luque, R., & Varma, R. S.: (Eds.). Sustainable Preparation of Metal Nanoparticles: Methods and Applications (Green Chemistry Series). Royal Society of Chemistry Vol. **19** (2012). https://doi.org/10.1039/9781849735469
54. Margarita, P., Salvador, G., Pablo, G., Ana M.: Costero Mesoporous Silica Nanoparticles in Chemical Detection: From Small Species to Large. Bio-Molecules Sensors. **22**(1) 261 (2022). https://doi.org/10.3390/s22010261
55. Sundus J., Amina B., G.: A Review on the Synthesis and Functionalization of Gold Nanoparticles as a Drug Delivery Vehicle. International Journal of Nanomedicine Volume. **15** 9823–9857 (2020). https://doi.org/10.2147/IJN.S279094

Chapter 3
Characterization of Gold Nanoparticles

Abstract Gold nanoparticles (AuNPs) possess distinct optical, chemical, and structural properties that are crucial for various nanotechnology applications. Their size, shape, purity, and dispersity govern their utility in biomedicine, catalysis, and electronics; therefore, precise characterization is essential. Transmission electron microscopy (TEM) and scanning electron microscopy (SEM) provide high-resolution imaging for determining size and morphology. Ultraviolet-visible (UV–Vis) spectroscopy is employed to determine the localized surface plasmon resonance (LSPR) of the nanoparticles, which is useful in evaluating their stability and aggregation behavior. Dynamic light scattering (DLS) measures the hydrodynamic diameter and zeta potential, making it suitable for assessing colloidal stability. X-ray diffraction (XRD) confirms crystallinity, while X-ray photoelectron spectroscopy (XPS) determines elemental composition. This chapter underscores the significance of accurate AuNP sizing and the critical role of these techniques in real-world applications. The advancement of these characterization methods will shape the future of nanotechnology, precision medicine, and sustainable material science.

3.1 Overview

Gold Nanoparticles (AuNPs) display fascinating and distinct behaviors when compared to their bulk and atomic counterparts. This is primirily due to key phenomena: quantum confinement effects, surface plasmon resonance, and a relatively higher surface-to -volume ratio [1]. These distinguishing characteristics render AuNPs highly useful across a range of applications in various industries, including healthcare, electronics, and catalysis. In this chapter, we aim to discuss in detail some of the current methods used for determining important characteristics of nanoparticles such as their size, shape, purity and dispersity. In addition, this chapter highlights the various techniques that have been applied to fabricated AuNPs and underscores the importance of characterization for their effective utilization in real-world applications.

A. C. Alguno et al., *Gold Nanoparticles*, Engineering Materials,
https://doi.org/10.1007/978-981-96-6565-5_3

The characterization techniques for AuNPs discussed in this chapter include Ultraviolet-visible (UV-Vis) spectroscopy, dynamic light scattering (DLS), transmission electron microscopy (TEM) and x-ray diffraction (XRD), as reported by [2]. These methods employed to determine the quality and size of the AuNPs and are elaborate upon in this section. They also address critical questions regarding the purity, size, and dispersion of the nanoparticles, which are of significant interest in the field of nanotechnology. Furthermore, UV-Vis spectroscopy requires optically active nanoparticles to interact with electromagnetic radiation, which is necessary to excite surface plasmon oscillations of the conduction electrons within the particles. UV–vis absorbance remains a highly effective tool for nanoparticle characterization. For asssessing nanoparticle size and distribution, DLS is widely utilized.

3.2 Techniques for Characterizing AuNPs

Analyzing AuNPs is essential for determining their physical, chemical, and optical properties, which dictate their performance across numerous applications, including biomedical science, catalytic processes, and electronic technologies. Scientists employ various analytical methods to evaluate essential parameters such as size, shape, crystallinity, and surface characteristics to achieve consistent and precise synthesis outcomes.

Transmission Electron Microscopy (TEM) and Scanning Electron Microscopy (SEM), both used in electron microscopy, allow extensive characterization of AuNP morphology [3, 25]. TEM enables high-resolution imaging that facilitates accurate determination of particle size and internal structure, whereas SEM provides complementary three-dimensional information on surface morphology, which is critical for understanding aggregation states and surface roughness. X-ray-based methods such as X-ray Diffraction (XRD) and X-ray Photoelectron Spectroscopy (XPS) are vital for identifying the crystal structure and evaluating the elemental composition of AuNPs [4, 5]. XRD confirms the crystalline nature of metalic nanoparticles, XPS detects impurities, chemical states, and the local environment of the elements in the nanoparticles.

The size and shape of AuNPs are monitored through localized surface plasmon resonance (LSPR), which is tracked using spectroscopic techniques such as Ultraviolet–Visible (UV–Vis) spectroscopy. This method is particularly effective for determining optical properties relevant to applications, such as sensing and imaging [6]. One of the key applications of AuNPs is analyzing their interactions with nearby functional groups through absorption techniques like Fourier-transform infrared (FTIR) and Raman spectroscopy, which reveal the behavior of AuNPs when integrated with surrounding materials. The hydrodynamic diameter and zeta potential are critical parameters that can be determined by dynamic light scattering (DLS). This technique is essential for asssessing the colloidal stability of nanoparticle dispersions. DLS provides a multiparametric profile of AuNPs in solution using time-resolved fluorescence anisotropy and ultrafast optical methods. AuNPs are colloidal solutions of gold particles produced via physical and chemical techniques.

In this chapter, we describe in detail the synthetic protocols of AuNPs and emphasize the role of stabilizing agents together with the necessity of well-controlled reaction conditions. This collection of characterization methods, which includes microscopy, spectroscopic, and electrochemical techniques, offers a comprehensive insights into the properties of AuNPs. Such enhancement provide critical information on size, shape, surface structure, and chemical composition, thereby contributing to their optimization for various applications and supporting that advancement of nanotechnology in both fundamental research and real- world applications.

3.2.1 Microscopy: Transmission Electron Microscopy (TEM), Scanning Electron Microscopy (SEM)

Two very important electron microscopy techniques that provide deeper insight into AuNPs characterization are TEM and SEM. TEM is well-known for its high-resolution imaging, enabling visualization of the internal structure of nanoparticles at a resolution of 1 nm or better [7]. This technique operates by transmitting a narrow electron beam through a thin sample, with contrast arising from interactions between the electrons and the sample. TEM is very effective in determining key characteristics such as particle size distribution, crystallography, lattice fringes, which reveal information about the atomic arrangement in the AuNPs. High-angle annular dark field imaging, a variant of TEM, extends its wide application by allowing distinction of core shell structures in functionalized AuNPs and by enabling the analysis of structural complexity in nanoparticles.

However, SEM provides a three-dimensional view of the nanoparticles and focuses on their surface morphology [8]. In SEM, the electron beams interacts with the surface of AuNPs, producing secondary electrons that generate high resolution images of the nanoparticle surfaces. This technique is particularly useful for analyzing larger aggregates and surface topology of AuNPs. Nevertheless, SEM has a lower resolution than TEM, it is well suited for examining bulk samples and mapping distribution of AuNPs on substrates. Furthermore, advanced SEM techniques such as Energy Dispersive Spectroscopy (EDS) enable elemental analysis, confirming the presence of gold and identify other element present in the sample.

Hence, both transmission and scanning electron microscopy yield high-resolution images that reveal the size, shape and dispersity of AuNPs and are well suited for detailed qualitative and quantitative analysis. Conventional imaging methods may not be capable of directly imaging colliodal suspensions, but these advanced techniques offer vital information regarding bulk AuNPs. Through precise electron interactions and sophisticated signal processing, these imaging techniques display the fine structure of AuNPs dispersion, including morphology, topography and elemental composition. In conclusion, TEM and SEM are complementary in nanoparticle characterization, providing insight into both internal structure and surface characteristics. This multifaceted approach enhances our understanding of AuNPs and strengthens applications in their potential applications in fields such as nanomedicine, electronics, and material science. Continued advancement of electron microscopy hold promise for deepening our comprehension of the nanoscale world and expanding opportunities for the development of AuNPs and related nanostructured systems.

3.2.2 *Spectroscopy: UV–Vis Spectroscopy, X-ray Diffraction (XRD)*

UV–Vis spectroscopy is employed to analyze the optical properties of AuNPs, as well as to obtain information on particle size, shape, and dispersion. According to Devika et al. [9], Ielo et al. [10], Localized surface plasmon resonance (LSPR) is one of the key detection criteria, manifesting in UV–Vis absorption spectra as an absorption peak between 500–600 nm. This peak is attributed to metallic nanoparticles and results from collective oscillation of conduction band electrons. It is highly sensitive to structural parameters. Aggregation, shape anisotropy, and surface modifications are known to induce shifts in the peak position and changes in intensity. The UV–visible extinction spectrum of AuNPs is also highly distinctive; the color of the solution may appear reddish violet, blue, or gray depending on the particle size and shape. Larger spherical particles exhibit a broad absorption band extending from visible region to the near infrared (NIR). Due to these optical features, AuNPs have found applications in imaging, therapy, and sensing.

X-ray diffraction is an important for characterizing the crystalline structure of AuNPs and serves as a valuable complement to UV–Vis spectroscopy. By analyzing the XRD patterns [11, 12], we can determine how nanoparticles are periodically arranged in the crystal lattice. XRD is commonly used to confirm the face-centered cubic (FCC) crystalline structure of gold and to estimate the crystallite size using the Scherrer equation. Moreover, secondary phase or impurities can be detected, ensuring that the synthesized nanoparticles are of high purity. The Bragg equation has been applied to explore the crystalline nature of the material, and the individual lattice planes of gold can be indexed accordingly. The absence of diffraction signals for certain samples suggests an amorphous nature, which is further supported by corresponding UV–Vis data.

3.3 Determining Physical Properties

The size, shape, and surface area of AuNPs are key physical properties that require optimization. These factors are critical as they influence the suitability of AuNPs for various applications [10, 13]. Techniques such as DLS enable rapid estimation of hydrodynamic diameters, while Atomic Force Microscopy (AFM) provides detailed information on nanoscale surface topology. These techiques also assist in ensuring that AuNPs are appropriately designed for specific biomedical applications, where stability and dispersibility are essential.

Essential characterizations include determining the mean particle size, polydispersity, zeta potential, and other intrinsic properties. UV–Vis spectroscopy is used to confirm the successful synthesis of nanoparticles, as evidenced by their LSPR features. The LSPR color depends on the particle size, shape, size distribution, and the dielectric constants of both the particles and the surrounding medium [14, 15]. Additionally, tensimetry is employed for characterizing smaller particles above 2 nm, as it reveals the details regarding the interaction between charged surfactants present at the nanoparticle surface. In terms of optical properties, the absorption peak for AuNPs larger than 2 nm typically appears in the visible region at around 520 nm. This absorption maximum shifts toward longer wavelenths (red shift) with increasing particle size and is accompanied by an increase in the intensity of the spectral response. Tonic analysis and fitting procedures can be used to extract particle size information directly from the UV–Vis spectra.

Another crucial parameter is the zeta potential, which measures electrokinetic stability. The AuNPs are typically negatively charged nanoparticles, and reported potential values generally ranged between −36 and −50 mV. This negative charge arises from the presence of hydroxyl and sulfate or sulfonate groups on the nanoparticles surface. Such electro-kinetic behavior is critical for ensuring colloidal stability, particularly in the context of medical applications involving AuNPs. These parameters are of significant importance in the rational design of AuNPs for targeted drug delivery systems.

3.3.1 Size, Shape, and Morphology

The properties of AuNPs are primarily attributed to their structural specificity, which differs significantly from that of bulk materials. These properties are highly sensitive to size, shape, and both intrinsic and extrinsic factors. This is because morphology is often influenced the synthesis of AuNPs, which greatly impacts their optical, electronic, and catalytic behavior [16, 17]. For instance, rod-shaped AuNPs exhibit different LSPR features compared to spherical ones, underscoring the need for precise morphological engineering.

Analytical techniques are essential for determining the morphology of AuNPs. Size and uniformity are typically assessed using TEM and SEM while XRD is employed to evaluate the crystallinity of nanoparticles as a function of size [18]. In addtion, AFM provides the three-dimensional surface profiles that offer further insight into surface topography.

The properties of AuNPs are size dependent and play a critical role in biomedical applications, particularly with respect to cellular interactions. Studies have demonstrated a quantitative relationship between the size and shape of nanoparticles and their cellular internalization, with these physical parameters being key determinants of uptake efficiency. It has been observed that certain morphologies facilitate enhanced internalization by increasing particle diameter, while others promote uptake through complex pathways that increase contact with the cell membrane.

Therefore, precise control over the size and shape of AuNPs is essential for their optimal use in various applications, including drug delivery and biomedical imaging. This chapter emphasizes the significance of morphology as a defining factor in cellular responses, which is vital for the rational design of gold bioconjugates for therapeutic use. Continued progress in synthesis and characterization techniques will support the advancement of AuNPs applications in critical areas of science and medicine.

3.3.2 Surface Plasmon Resonance (SPR)

Surface plasmon resonance (SPR) is a surface property of AuNPs that arises from the resonance of conduction electrons in the presence of light. This phenomenon is also responsible for the absorption band associated with the formation of coherent charge density waves at the metal surface, which accounts for the characteristics reddish color of dispersed AuNPs due to their SPR behavior [19]. The optical properties of AuNPs, particularly valued for biosensing applications, are strongly influenced by the size, shape and type of particles, all of which significatly affect the optical response during interactions.

SPR can be readily investigated using UV–Vis spectroscopy, where variations in peak positions are correlated with the size, shape and degree of aggregation of the nanoparticles. Owing to the high sensitivity of SPR to environmental changes, it is widely utilized to target detection. Any shift in the absorption peak is attributed to the specific adsorption of biomolecules on the nanoparticle surface. This property has enabled the design of multifunctional nanosensors capable of directly detecting clinically relevant biomarkers and ions with high efficiency. Furthermore, techniques such as Surface-Enhanced Raman Scattering (SERS) exploit the SPR effect to detect biomolecules with ultrasensitivity by amplifying the emission signal through the coupling of light with molecular vibrations. The size, shape, or interaction of AuNPs with surrounding molecular environments can significantly influence their optical behavior, including signal quenching, enhancement, or spectral shifts.

In sensing applications, such as the detection of Hg^{2+} and Pb^{2+} ions, AuNPs exhibit photoinduced variations in their near-infrared absorption, which refelect ion concentration by inhibiting nanoparticle aggregation in acquoes solutions. The multifucntional capabilities of AuNPs driven by SPR provide a robust foundations for advancing nanosensor technology, with broad applications aross environment monitoring and clinical diagnostics.

3.4 Surface Functionalization and Stability

The stability and application of AuNPs in various fields such as biomedicine and environmental sensing depend significantly on the surface functionalization of the nanoparticles. This process involves the attachment of specific molecules to the nanoparticle surface to prevent aggregation, enhance biocompatibility, and modulate interactions with the target analytes [20]. Surface coatings play a critical role in maintaining AuNP stability. Common stabilizers such as citrate ions, polyethylene glycol (PEG), and thiol groups operate through different mechanisms, For instance, citrate ions induce electrostatic stabilization, while thiol-based molecules form strong bonds with the gold surfaces, leading to the formation of stable self-assembled monolayers. These coatings also establish a hydration layer that minimizes non-specific interactions, thereby improving the biological half-life of the nanoparticles.

Functionalization also enables the introduction of specific functional groups, allowing AuNPs to interact selectively with chemical or biological targets. For example, the conjugation of biomolecules such as antibodies, peptides, or nucleic acids transforms functionalized AuNPs into effective vehicles for targeted drug delivery or use in biosensing applications [21]. Importantly, the stability of these functionalized nanoparticles is maintained under various conditions including changes in pH, ionic strength, and temperature, which is essential for their use in diagnostic and therapeutic contexts.

Polymers are commonly used in functionalization due to their ease of modification. They can facilitate electrostatic interactions that minimizes the presence of unbound ions during nanoparticle synthesis and encourage the formation of well-structured aggregates. Notable advancements include the successful functionalization of PEG through thiol-ene reactions and derivatization of polyvinylpyrrolidone (PVP) into thiol-terminated species. The interaction between AuNPs and biomolecules such as bovine serum albumin (BSA) has also been shown to alter the optical properties of the nanoparticles upon functionalization. Additionally, AuNPs have demonstrated utility in environmental applications, where functionalized polymers are also capture hazardous substances through specific interactions in sensing technologies. This highlights the versatile applicability of functionalized AuNPs across a range of sensing platforms.

3.4.1 Techniques for Surface Modification

Surface modification is crucial in the design of AuNPs for specific applications and for enhancing their physical properties. This section focuses on the SPR of AuNPs, which is a surface property arising from the resonance of conduction electrons in the presence of light. This phenomenon is also responsible for the absorption band that leads to the formation of coherent charge density waves at the metal surface. This, in turn, accounts for the characteristic reddish color of dispersed AuNPs due to their SPR behavior [19]. The optical properties of AuNPs, widely recognized for their role in biosensing applications, are strongly influenced by the particle size, shape and type, all of which significantly affect the optical responsed during interaction processess.

The size, shape and aggregation state related to SPR can be effectively investigated using UV–Vis spectroscopy. Variations in peak positions are highly sensitive to environment changes and are thus employed for target detection. Any shift in the absorption peak is typically attributed to the specific adsorption of biomolecules on the surface of the nanoparticles. More advanced techniques, including dual-ligand systems and layer-by-layer strategies, have been developed to enhance both biocompatibility and functional performance. Nanoparticle stability and cell functionality in complex biosystems. These approaches collectively address critical issues such as maintaining nanoparticle stability and ensuring cellular compatibility in complex biological environments.

In addition, ligand exchange methods, such as place-exchange reactions and protective agent-assisted synthesis, allow the selective modification of AuNPs. Place-exchange processes may be constrained by reaction kinetics and often require high concentrations of the exchange species. To address this, physical stimuli can be applied to accelerate the exchange process. Alternatively, direct synthesis of functionalized nanoparticles in the presence of ligands can yield particles with tailored protective shells. However, achieving water solubility remains a significant challenge for certain applications. Despite advances in functionalization techniques, the limited availability of methods capable of simultaneously incorporating multiple ligands highlights the need for innovative strategies to overcome current limitations. Expanding sucha capabilities is essential in improving the versatility and applicability of functionalized AuNPs in various fields.

3.4.2 Stability in Different Media (Aqueous vs. Organic)

The stability of AuNPs in both aqueous and organic media is intrinsically linked to their surface properties and the surrounding environment. Electrostatic stabilization plays a key role in aqueous solutions, where negatively charged ligands such as citrate effectively prevent aggregation by electrostatic repulsion [22, 26] .

However, high ionic strength or extreme pH conditions can destabilize this interaction, leading to nanoparticle aggregation. To adress this limitation, PEGylation is commonly applied employed to enhance colloidal stability and biocompatibility, especially in physiological environments.

On the other hand, the stability of AuNPs in organic solvents is largely determined by the compatibilty of the ligand shell with solvent polarity. For instance, alkylthiol hydrophobic ligands provide steric stabilization in nonpolar solvents such as chloroform and toluene [27]. Phase transfer techniques enable functionalization of AuNPs with organic ligands, facilitating their stabilization of nanoparticles in either aqueous or organic media and broadening their utility in catalytic and other applications. Importantly, AuNPs exhibit vivid coloration and disperse well in several organic solvents, such as low-polar tetrahydrofuran (THF) and polar methanol, which offer long-term storage stability in storage. AuNPs appear to be insoluble in solvents such as Ethyl acetate and chloroform when hydrophilic content in high, due to solubilityy gradients. Solubility in organic solvent follows the decreasing order of functional groups: carboxylic acid, amine, and hydroxyl.

Although AuNPs generally exhibit higher stability in aqueous media compared to organic environments, their stability can be further enhanced by surface coating with functional groups such as primary amines and carboxylic acids. The careful selection of these functional groups supports self-assembly processes, inducing strong dichroic effects, and enables tunable plasmonic absorption. Additionally, AuNPs shows exhibit phase transformations driven by gold–thiol interactions. The ability to finely control surface properties through tailored reaction conditions and reagent selection emphasizes the versatility of AuNPs for advanced applications in nanotechnology.

3.5 Role of Characterization in Application Development

Understanding the properties of AuNPs is essential when designing them for specific applications in biosensing, catalysis, and drug delivery. Primary characterization methods include TEM and DLS, which provide the valuable information on particle size, shape, and size distribution, parameters that are crucial for both optical and catalytic performance [23]. The size and surface chemistry of AuNPs can also be evaluated using SPR, typically analyzed by UV–Vis spectroscopy.

Verification of functionalization is very important to assure that the desired bioactive agents are successfully incorporated in the system. This is typically achieved using FTIR and XPS [28]. These techniques offer insights into surface chemistry and confirm the attachment of functional ligands, which enhances the selectivity and the sensitivity of AuNP-based applications, especially in biosensing. The effectiveness of AuNPs also depends on the appropriate selection of reducing and capping agents, which determines the resulting surface chemistry. Chemical reduction methods are commonly employed for synthesizing nanoparticles with varied size and shape that are tailored for specific functionalities. UV–Visible spectroscopy is routinely used to assess particle size, shape, concentration and optical density, while DLS provides information on agglomeration behavior and dispersion stability.

Furthermore, SEM offer further insight into particle morphology when used in conjunction with TEM, and DSC employed for thermal analysis. In recent years, there has been a notable shift toward green synthesis approach, which utilize natural substances as reducing and capping agents. These environmentally friendly methods minimize ecological impact and often results in stable nanoparticle suspensions with extended shelf life. Accordingly, this section provides synergistic insights into the characterization techniques and synthesis strategies that underpin the effective use of AuNPs across diverse scientific disciplines.

3.5.1 Understanding Properties for Specific Applications (Biosensors, Catalysis)

AuNPs are widely employed in biosensing and catalysis due to their optical, electronic, and catalytic properties, which are strongly dependent on their size and shape, parameters influenced by the conditions of their synthesis [17]. The surface plasmon resonance of AuNPs is characteristic feature that makes them suitable for highly sensitive biomolecule detection, utilizing colorimetric or fluorescence-based dendrimers systems. These systems respond to changes in the dielectric environment surrounding the nanoparticles. Such changes are exploited for the detection of toxins, proteins, microorganisms, and even DNA mutations that can lead to genetic damage.

AuNPs are known for their surface reactivity and electronic structure, both of which enhance reaction rates such as carbon monoxide oxidation and hydrogenation [24]. Catalysts incorporating AuNPs with high surface area have shown an increased efficiency in these reactions. Advances in AuNP synthesis have led to the development of gold-based catalysts that play a significant role in heterogeneous catalysis and are now applied to a wide range of chemical transformations.

Recent strategies have introduced materials such as sodium dodecylbenzene sulfonate to improve the selectivity of AuNPs when used as catalytic supports. This surfactant can forms self-assembled bipolar micelles that optimize the interface between AuNPs and the substrate. Furthermore, the nonlinear optical properties of small AuNPs open new avenues for signal generation through nonlinear photolysis. The integration of gold nanoparticles in biosensors and catalytic systems seems holds great promise for addressing key challenges in biochemical detection and catalysis, contributing significantly to the ongoing advancement of nanotechnology.

References

1. Philip, A., Kumar, A.R.: The performance enhancement of surface plasmon resonance optical sensors using nanomaterials: a review. Coordination Chem. Rev. (2022).https://doi.org/10.1016/j.ccr.2022.214424
2. Arshad, F., Hassan, I.U., Naikoo, G.A.: Characterization of nanomaterials. Nanotechnol. Herbal Med. (2023). https://doi.org/10.1016/B978-0-323-99527-6.00005-7
3. Lyu, Z., Yao, L., Chen, W., Kalutantirige, F.C., Chen, Q.: Electron microscopy studies of soft nanomaterials. Chem. Rev. **123**(7), 4051–4145 (2023). https://doi.org/10.1021/acs.chemrev.2c00461
4. Schmutzler, O., Graf, S., Behm, N., Mansour, W.Y., Blumendorf, F., Staufer, T.,... Grüner, F.: X-ray fluorescence uptake measurement of functionalized gold nanoparticles in tumor cell microsamples. Int. J. Molec. Sci. **22**(7), 3691 (2021). https://doi.org/10.3390/ijms22073691
5. Whitehead, C.B., Finke, R.G.: Particle formation mechanisms supported by in situ synchrotron XAFS and SAXS studies: a review of metal, metal-oxide, semiconductor and selected other…. Mater. Adv. (2021). https://doi.org/10.1039/d1ma00222h
6. Pashkov, D.M., Guda, A.A., Kirichkov, M.V., Guda, S.A., Martini, A., Soldatov, S.A., Soldatov, A.V.: Quantitative analysis of the UV–vis spectra for gold nanoparticles powered by supervised machine learning. J. Phys. Chem. C **125**(16), 8656–8666 (2021). https://doi.org/10.1021/acs.jpcc.0c10680
7. Chao, H.Y., Venkatraman, K., Moniri, S., Jiang, Y., Tang, X., Dai, S.,... Chi, M.: In situ and emerging transmission electron microscopy for catalysis research. Chem. Rev. **123**(13), 8347–8394 (2023). https://doi.org/10.1021/acs.chemrev.2c00880.
8. Cid-Mejías, A., Alonso-Calvo, R., Gavilán, H., Crespo, J., Maojo, V.: A deep learning approach using synthetic images for segmenting and estimating 3D orientation of nanoparticles in EM images. Comput. Methods Programs Biomed. **202**, 105958 (2021). https://doi.org/10.1016/j.cmpb.2021.105958
9. Devika, V., Sreelekshmi, P.J., Pai, R.G., Sadanandan, S.: Spectroscopic and microscopic characterizations of gold nanoparticles. In: Gold Nanoparticles, Nanomaterials and Nanocomposites, pp. 201–229. Elsevier (2025). https://doi.org/10.1016/B978-0-443-15897-1.00019-4
10. Ielo, I., Rando, G., Giacobello, F., Sfameni, S., Castellano, A., Galletta, M.,... Plutino, M.R.: Synthesis, chemical–physical characterization, and biomedical applications of functional gold nanoparticles: a review. Molecules **26**(19), 5823 (2021).https://doi.org/10.3390/molecules26195823
11. Gruene, T., Holstein, J.J., Clever, G.H., Keppler, B.: Establishing electron diffraction in chemical crystallography. Nat. Rev. Chem. **5**(9), 660–668 (2021). https://doi.org/10.1038/s41570-021-00302-4
12. Zhu, Y., Kuo, T.R., Li, Y.H., Qi, M.Y., Chen, G., Wang, J.,... Chen, H.M.: Emerging dynamic structure of electrocatalysts unveiled by in situ X-ray diffraction/absorption spectroscopy. Energy Environ. Sci. **14**(4), 1928–1958 (2021).https://doi.org/10.1039/D0EE03903A
13. Gu, X., Xu, Z., Gu, L., Xu, H., Han, F., Chen, B., Pan, X.: Preparation and antibacterial properties of gold nanoparticles: a review. Environ. Chem. Lett. **19**, 167–187 (2021). https://doi.org/10.1007/s10311-020-01071-0
14. Khurana, K., Jaggi, N.: Localized surface plasmonic properties of Au and Ag nanoparticles for sensors: a review. Plasmonics (2021). https://doi.org/10.1007/s11468-021-01381-1
15. Xu, T., Geng, Z.: Strategies to improve performances of LSPR biosensing: structure, materials, and interface modification. Biosensors and Bioelectronics (2021).https://doi.org/10.1016/j.bios.2020.112850
16. Kim, J.M., Lee, C., Lee, Y., Lee, J., Park, S.J., Park, S., Nam, J.M.: Synthesis, assembly, optical properties, and sensing applications of plasmonic gap nanostructures. Adv. Mater. **33**(46), 2006966 (2021). https://doi.org/10.1002/adma.202006966
17. Sarfraz, N., Khan, I.: Plasmonic gold nanoparticles (AuNPs): properties, synthesis and their advanced energy, environmental and biomedical applications. Chem. Asian J. (2021). https://doi.org/10.1002/asia.202001202

18. Tan, H., Huang, Z.C., Zhu, S.R., He, L., Fang, X.: An image-based approach for quantitative assessment of uniformity in particle distribution of noise reduction material. Microsc. Res. Tech. **84**(9), 1924–1935 (2021). https://doi.org/10.1002/jemt.23748
19. Ai, K., Huang, J., Xiao, Z., Yang, Y., Bai, Y., Peng, J.: Localized surface plasmon resonance properties and biomedical applications of copper selenide nanomaterials. Mater. Today Chem. (2021).https://doi.org/10.1016/j.mtchem.2020.100402
20. Ahmad, F., Salem-Bekhit, M.M., Khan, F., Alshehri, S., Khan, A., Ghoneim, M.M.,... Elbagory, I.: Unique properties of surface-functionalized nanoparticles for bio-application: functionalization mechanisms and importance in application. Nanomaterials **12**(8), 1333 (2022). https://doi.org/10.3390/nano12081333
21. Taneja, P., Sharma, S., Sinha, V.B., Yadav, A.K.: Advancement of nanoscience in development of conjugated drugs for enhanced disease prevention. Life Sci. (2021).https://doi.org/10.1016/j.lfs.2020.118859
22. Lim, S.H., Wong, T.W., Tay, W.X.: Overcoming colloidal nanoparticle aggregation in biological milieu for cancer therapeutic delivery: perspectives of materials and particle design. Adv. Colloid Interface Sci. (2024).https://doi.org/10.1016/j.cis.2024.103094
23. Lin, Y., Zhou, M., Tai, X., Li, H., Han, X., Yu, J.: Analytical transmission electron microscopy for emerging advanced materials. Matter (2021).https://doi.org/10.1016/j.matt.2021.05.005
24. Mala Wani, I.A., Jain, S.K., Khan, H., Kalam, A., Ahmad, T.: Gold nanoparticles as efficient catalysts in organic transformations. Curr. Pharm. Biotechnol. **22**(6), 724–732 (2021). https://doi.org/10.2174/1389201022666210218195205
25. Manuela, M.: Transmission Electron Microscopy as a Powerful Tool to Investigate the Interaction of Nanoparticles with Subcellular Structures. International Journal of Molecular Sciences. **22**(23) 12789–10 (2021).https://doi.org/10.3390/ijms222312789
26. Kaustav, B., Bhagavatula L. V., P.: Surface functionalization of inorganic nanoparticles with ligands: a necessary step for their utility. Chemical Society Reviews. **52**(8) 2573–2595 (2023). https://doi.org/10.1039/D1CS00876E
27. Yue, L., Min, Z., Xiaolin, T., Hangfei, L., Xiao, H., Jiaguo, Y.: Analytical transmission electron microscopy for emerging. advanced materials Matter. **4**(7) 2309–2339 (2021).https://doi.org/10.1016/j.matt.2021.05.005
28. Boobas, S., Jayaprakash, J., Ranjith, R., Priyadharsan, A., Krishnakumar, V.: Visible light photocatalytic activity of tungsten and fluorine codoped TiO2. nanoparticle for an efficient dye degradation Ionics. **25**(2) 773–784 (2019). https://doi.org/10.1007/s11581-018-2628-x

Chapter 4
Green Synthesis of Gold Nanoparticles Using Plant Extracts

Abstract As a "green" method for the synthesis of gold nanoparticles (AuNPs) plant extracts can be used instead of synthtic reagents. Thus, the use of natural flavonoids, alkaloids and phenolics as reducing and stabilizing agents is preferable to toxic chemicals. The synthesis process consists of reduction, nucleation and stabilization, and the size and shape of the nanoparticles influenced by pH, temperature and extract concentration. Many plants such as neem, tulsi and aloe vera have been found to be effective in the production of biocompatible AuNPs, which have found application in biomedicine, catalysis, and environmental monitoring and remediation. Reaction conditions are optimized to improve nanoparticle stability and functionality. Plant cell fermentation enhances control over the synthesis process and reduces the impact on the environment. Despite the problems of standardization and industrialization, plant-mediated synthesis of nanoparticles is a sustainable approach. This chapter highlights the mechanisms, advantages, and possibilities of plant-based AuNP synthesis in advancing green chemistry, precision medicine, and environmental friendliness.

4.1 Overview

A new environmentally friendly method for the synthesis of gold nanoparticles (AuNPs) is based on the use of plant extracts as reducing and stabilizing agents. This alternative method is considerably safer for the environment because it does not involve the use of toxic chemicals commonly employed in the conventional fabrication of nanoparticles. The use of plant-based methods is not only economically advantageous but can also be extended to large-scale applications in various fields, including medicine, environmental remediation and food technology. The AuNPs prepared by this green method have favorable optical and chemical properties, including high biocompatibility and stability, which make them suitable for use in diagnostic assays, catalytic systems, and drug delivery platforms. This method of nanoparticle synthesis is consistent with international goals that promote green chemistry and sustainable development.

A. C. Alguno et al., *Gold Nanoparticles*, Engineering Materials,
https://doi.org/10.1007/978-981-96-6565-5_4

4.2 Mechanism of Plant-Mediated Synthesis

One of the most promising emerging applications of nanotechnology is the plant-mediated synthesis of AuNPs, offering an innovative, green, efficient and versatile route for the synthesis of nanoparticles that is scientifically relevant and environementally sustainable. This unique process occurs in three crucial steps, reduction, nucleation, and stabilization, each dependent on a diverse array of phytochemicals present in the plant extract [1, 2]. It begins with the reduction of the gold-ion solution using biomolecules inherent to plant sources, such as polyphenols, flavonoids, alkaloids, and terpenoids, as green reducing agents (Fig. 4.1).

These biomolecules assist in electron transfer leading to the reduction of Au^{3+} to Au^0 metallic gold, thereby initiating nanoparticle formation. During the initial phase of reduction, gold atoms absorbed into biomelecular surfaces collide and cluster together to form small nuclei, which serve as the starting point for the further nucleation and growth of gold crystals. At this stage, factors such as pH, temperature, and extract concentration play a crucial role in determining the size and shape of the nanoparticles [2, 3]. An alkaline environment facilitates the formation of uniform nanoparticles, while elevated temperatures enhances reaction kinetics, promoting accelerated growth. The last stage, stabilization, occurs when plant-based biomolecules coat the nanoparticles, inhibiting agglomeration and simultaneously providing stability in colloidal media.

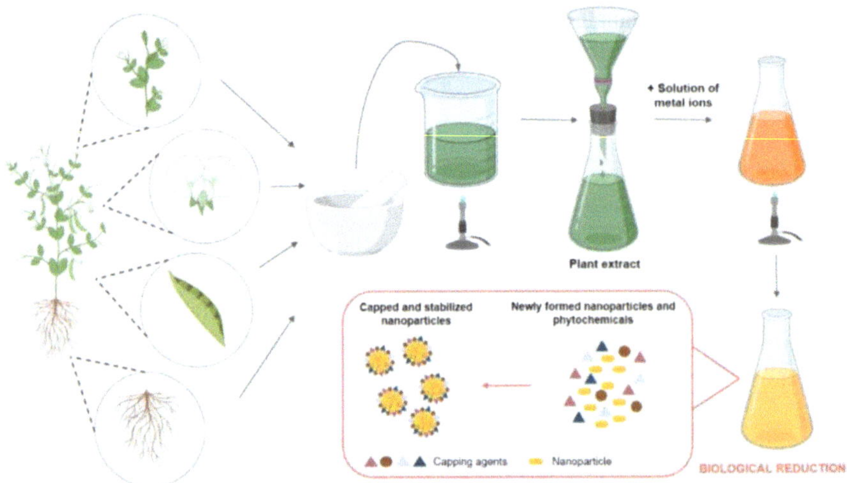

Fig. 4.1 Schematic representation of the green synthesis of nanoparticles using plant extracts. Various plant parts (leaves, flowers, seeds, roots) are processed to obtain phytochemical-rich extracts, which are then mixed with a metal ion solution. Adapted from [2], Fig. 4.1, https://doi.org/10.3390/molecules26040844. Licensed under CC BY 4.0 (https://creativecommons.org/licenses/by/4.0/)

In addition to enhancing the stability, these monolayers give specific surface properties to the gold nanoparticles that make them useful for several applications in biomedicine, catalysis, and optics. Plant-mediated synthesis has the potential to revolutionize the field because of its low-cost and small-scale feasibility over conventional methods, avoiding toxic chemicals and expensive equipment. The abundant diversity of plant sources, including leaves and roots, provides a reservoir of phytochemicals for the fine-tuning of the physicochemical properties of nanoparticles, thereby transforming nanoparticle synthesis. This plant-mediated synthetic paradigm symbolizes a significant milestone in sustainability in nanotechnology and opportunities for the economically viable synthesis of catalytic-grade high-purity AuNPs may be fruitful for hundreds of next-generation applications.

4.2.1 Role of Phytochemicals: Alkaloids, Flavonoids, and Phenolic Compounds

The synthesis of AuNPs from plant-based sources reflects how green chemistry can converge with nanotechnology by utilizing the bioactive constituents of plant extracts in nanoparticle formation. The process gives primacy to phytochemicals, which are essential in the synthesis and stabilization of nanoparticles by the reduction of metal ions, endowing them with functionalities and colloidal stability [4, 5]. Among these phytochemicals, heterocyclic nitrogen-containing alkaloids are of utmost importance in reducing gold ions (Au^{3+}) to Au^0 by donating electrons.

The structural details of these alkaloids also act as stabilizers of the nanoparticles, contributing to nanoscale structural integrity and dispersion stability. These activities are attributed to a class of phytochemical compound called flavonoid which have been widely reported as responsible for AuNPs synthesis from numerous plants. In addition to being capping agents that inhibit agglomeration and extend shelf life [6], their hydroxyl groups also act as reducing agents during the controlled synthesis of the nanoparticles, and they regulate the size and shape of the nanoparticles. In addition, tannins and catechins are phenolic compounds that facilitate the synthesis by creating dense protective layers, which allow a homogeneous dispersion of the nanoparticles in aqueous media. Due to their prevalence, biocompatibility and low toxicity, these phytochemicals are of utmost significance in biomedicine and environmental applications. Consistent with this, the synergistic activity of alkaloids, flavonoids, and phenolic compounds in AuNPs generation further points to a great promise of these plant-derived compounds in the synthesis of AuNPs and other forms of nanomaterials in an eco-friendly manner.

4.2.2 Case Studies of Green Synthesis Using Plants

Green synthesis of AuNPs employing plant extracts as reducing agents is a novel and promising approach in the field of sustainable nanotechnology. This eco-sustainable approach utilizes plants as natural bioreactors to reduce, cap, and stabilize nanoparticles through their intrinsic bioactive compounds. However, as the following case studies illustrate, this approach ensures tremendous variety and exceptional efficacy.

A good plant to demonstrate the potential of this technique is neem (*Azadirachta indica*), a medicinal plant native to the Indian subcontinent. The extracts of neem contain alkaloids and flavonoids, which facilitate the formation of small, highly stable, spherical AuNPs and improve the biocompatibility of the nanoparticles, thereby enhancing their suitability for therapeutic applications [7]. For instance, tulsi (*Ocimum sanctum*), a plant native to South Asia, contains phenolic compounds and flavonoids with reducing and capping activity. The AuNPs obtained using these extracts are antimicrobially active and can find use in biomedical applications [8].

Among cultivated plants, aloe vera is commonly grown for its health and cosmetic uses, which contributes polysaccharides and phenolic acids to the nanoparticle formation process. The AuNPs obtained exhibit adjustable optical properties and strong surface plasmon resonance and thus can be used in advanced biosensing technologies [9]. *Eucalyptus globulus* is also a fast-growing species and is popularly used in Australia and other parts of the world for its polyphenols that help accelerate the synthesis process for the detection of heavy metal pollutants, roles that are relevant to the environmental applications of these nanoparticles [10].

Another important case is *Euphorbia fischeriana* because of its alkaloid and phenolic acid content, produces small, monodisperse and stable AuNPs that are suitable for catalytic and biomedical applications [11]. Furthermore, *Punica granatum* (pomegranate), a Mediterranean symbol of fertility, yields biocompatible AuNPs with distinctive optical features thanks to its polyphenols and tannins. The application of these nanoparticles is possible for drug delivery and biosensing [12]. Lastly, the alkaloids and flavonoids from *Carica papaya* allow for the preparation of plasmonic nanoparticles that exhibit good dispersion characteristics and are mainly utilized in environmental analyses for the detection of heavy metals [13].

These illustrative case studies demonstrate the great potential of the green synthesis approach. Thus, the variation in biochemical content among different plants can be exploited to control the size and shape of nanoparticles, thereby tailoring them for a wide range of applications in medicine and environmental science. This sustainable methodology also avoids the use of hazardous chemicals and is therefore in harmony with the sustainable development goals outlined by global frameworks. Therefore, it can be stated that this method will play an essential role in the future of nanotechnology (Table 4.1).

Table 4.1 Common plants used in the green synthesis of gold nanoparticles (AuNPs), highlighting their key biocompounds and resulting nanoparticle morphologies.

Common name	Scientific name	Biocompounds present	AuNP morphologies	References
Neem	*Azadirachta indica*	Alkaloids, flavonoids	Spherical, excellent size control, stable	[7]
Tulsi (Holy Basil)	*Ocimum sanctum*	Phenolic compounds, flavonoids	Spherical, antimicrobial	[8]
Aloe Vera	*Aloe vera*	Polysaccharides, phenolic acids	Adjustable optical properties, spherical	[9]
Eucalyptus	*Eucalyptus globulus*	Polyphenols	Spherical, sensitive to pollutant detection	[10]
Fischer Spurge	*Euphorbia fischeriana*	Alkaloids, phenolic acids	Uniform size, stable spherical	[11]
Pomegranate	*Punica granatum*	Polyphenols, tannins	Uniform, biocompatible spherical	[12]
Papaya	*Carica papaya*	Alkaloids, flavonoids	Excellent dispersion, plasmonic spherical	[13]

4.3 Optimization of Reaction Conditions

It is critically important to optimize the reaction conditions for the green synthesis of AuNPs, as it affects the size, shape and stability of the nanoparticles. Important factors include temperature, pH, reducing agent concentration and reaction time, all of which need to be systematically fine tuned for reproducibility [14, 15]. As an example, the formation of nanoparticles can be accelerated at higher temperatures because the rate of reduction of gold ions increases with temperature, but higher temperatures may also lead to uncontrolled growth and particle aggregation of particles. Likewise, the ionization of phytochemicals and their adhesion to the metal ions affect the formation of the nanoparticle morphology is affected by pH [16]. The increased concentration of plant extracts provides more reducing agents that actively participate in the reaction, resulting in smaller and uniform sized particles formation. Therefore, it is essential to recognize and regulate this interaction to ensure the successful generation of AuNPs suitable for application in catalysis, sensing, and biomedicine.

4.3.1 Effect of Plant Extract Concentration, Metal Ion Concentration, pH, and Temperature on AuNPs Formation

The synthesis of AuNPs depends on many factors that are interlinked, and plant extract concentration is one of the most important. This concentration determines the availability of important bioactive compounds, such as flavonoids, alkaloids, and phenolic acids, which act as both reducing and capping agents in the formation of nanoparticles. Overall, the concentration of the extract is directly proportional to the size and uniformity of the nanoparticles, since there are more stabilizing molecules available. At high concentrations however, particles may aggregate because an excess of reducing agents that exceed the stabilizing properties of the biomolecules.

Simultaneously, the density of the gold ions must be taken into account; in excess, the ions may aggregate or merge, since the exaggerated consumption of the metal surpasses the stabilizing power of the plant compounds. This dynamic is mediated by the pH of the reaction medium, which influences phytochemical ionization and gold ion reduction potential [17]. This allows for a faster reduction process under alkaline conditions, giving rise to small and uniformly shaped nanoparticles, whereas, under acidic conditions, the reduction is slower and can lead to nanoparticles with irregular morphologies. Moreover, temperature plays a critically important role, since high temperature raises the speed of gold ions reduction, but may compromise the stability of nanoparticles if too high. As a result, careful optimization of these parameters is critical for the successful green synthesis of AuNPs that are tailored to the demands of their intended applications.

4.4 Green Synthesis of AuNPs Using Cultivated Plant Cells

Sustainable synthesis using cultured plant cells has been realized for the green production of AuNPs, highlighting a precise combination of targeted control and eco-friendliness. Conventional techniques are challenging in terms of reproducibility and consistency due to the use of whole-plant extracts; however, these cell systems operate in a well-controlled environment [18]. They utilize a platform that exploits the plant cell s' natural biosynthetic pathways, which generate a wide array of secondary metabolites (including flavonoids, phenolic acids, and alkaloids) as bioactive agents allowing green conversion of Au^{3+} ions into stable nanocrystals.

Cultivated plant cells are significant because they can reduce the variability that is common in traditional extraction methods. Factors such as soil quality, climate conditions, and plant age are well-documented sources of variability in phytochemical content as well as nanoparticle properties. In contrast, cell cultures provide a consistent, high-yield source of biomolecules that enable control over size, morphology, and nanoparticle stability [19]. These characteristics are crucial in biomedical imaging, targeted drug delivery, and biosensing applications. Furthermore, this method is sustainable because it does not require extensive harvesting from plants, thereby avoiding adverse effects on the environment, and optimizing the use of resources. This approach not only demonstrate the potential of plant cell cultures for the development of advanced nanomaterials but also strengthens the concept of green nanotechnology, and defines the directions for the sustainable development of the field.

4.4.1 Callus Cultures and Cell Suspension Techniques

The callus cultures and cell suspension methods are presently advanced methods in the green synthesis of AuNPs with good accuracy, high throughput, and eco-friendliness. Callus cultures are the proliferation of undifferentiated plant cells on solid media wherein auxin and cytokinin type hormones induce the production of callus tissue [20]. These are metabolically active cells that contain bioactive compounds such as phenolics and flavonoids, which reduce Au^{3+} ions and stabilize the nanoparticles. This is because the environment in callus cultures is well controlled, leading to the production of nanoparticles of the desired size and shape.

However, cell suspension cultures are considered superior, as they engage plant cells in liquid media, on a platform that provides continuous stirring, thereby providing more contact between metabolites and metal ions, and faster formation of nanoparticles with better yields [21]. This medium allows for precise management of parameters such as pH, and temperature, optimizing nanoparticle characteristics such as stability, and morphology. Moreover, the use of elicitors, including jasmonic, and salicylic acid enhances the production of secondary metabolites, which further enhance the yield.

These techniques collectively contribute to enhancements in green nanotechnology, addressing classic variability problems, while also ensuring environmental sustainability. All of them follow the principles of green chemistry to facilitate small-scale, environmentally friendly production of nanoparticles for biomedical, industrial, and environmental purposes.

4.5 Pros and Cons of Microbial Synthesis

The synthesis of AuNPs by the microbial method has been reported to be a novel, ecofriendly method of synthesis, which uses the metabolic activities of microorganisms such as bacteria, fungi, and algae in the synthesis of the nanoparticles. This process is in compliance with the green chemistry guidelines, as it does not involve the use of toxic chemicals and harsh conditions, thus making the environment more sustainable. These microorganisms act as biological nanofactories that gold ions (Au^{3+}) to biocompatible metallic gold (Au^0) under ambient conditions, and therefore have biomedical applications potential. The second benefit of the microbially assisted synthesis method is that it can be scaled up with decreased production cost, allowing high yield that is economically viable for industries [8].

However, there are several problems that inhibit their widespread application. The variability of microbial behavior causes changes in the size and shape of the nanoparticles, and this is critical in precision-based applications [20]. Furthermore, there is a risk of contamination during microbial cultivation, which may affect the purity of the nanoparticles, along with the requirement for growth conditions that may be more demanding compared to those used in conventional chemical synthesis. Hence, advanced approaches must be implemented to overcome these limitations, enhance the reproducibility and control, and pave the way for using microbial synthesis as a sustainable and high-quality approach to the production of AuNPs.

4.5.1 Advantages: High Yield and Eco-Friendliness

An environmentally friendly method of synthesizing AuNPs through a microbial approach with high yield is reported, and it can be suggested as a potential alternative to the conventional chemical synthesis methods. This process takes advantage of the intrinsic ability of microorganisms to reduce metal ions into nanoparticles, often at ambient temperature and pressure, thereby reducing the energy input and material consumption. Furthermore, microbes produce naturally compatible capping agents such as proteins and polysaccharides, that facilitate the capping of the nanoparticles and make them more stable and active without the need for external stabilizers.

Fusarium oxysporum, a fungus, is known to reduce gold ions while simultaneously producing these capping agents, resulting in a relatively simple process [22]. Furthermore, the development of genetic engineering makes it possible to control microbial systems, enabling the synthesis of nanoparticles with desired size, shape, and properties for their maximum use in diagnostics, catalysis, and therapeutics. Microbial synthesis, therefore, replaces hazardous chemical reagents with biological systems, minimizing pollution and, thereby, embracing green chemistry in the development of nanotechnology.

4.5.2 Limitations: Contamination Risk and Purification Challenges

Microbial-based biosynthesis of AuNPs has many advantages, but it is also associated with several factors that need to be addressed. One of the major challenges is the risk of contamination, since microbial cultures are easily contaminated by foreign organisms, which compromise the reproducibility and purity of the nanoparticles obtained [23]. It is also important to maintain sterility during the synthesis, which renders the process more complex and costly. Purification is another fundamental challenge, as biogenic synthesis usually produces an impure mixture of nanoparticles associated with biological residues, including proteins, polysaccharides, cellular debris among others [24]. This separation of by-products requires sophisticated purification methods, such as centrifugation, filtration, or chromatography, that can be time-intensive and may lead to lower yield of the nanoparticle product, or cause a modification in its attributes.

Moreover, variability between microbial strains, as well as growth conditions and metabolic pathways, makes stable nanoparticle traits difficult to achieve, and industrial scale-up a challenge. Nonetheless, the field has not stagnated. There is active work underway in engineering microbes with better control of the synthesis, and purification steps, through new variants of the process, thereby keeping microbial synthesis an emerging powerhouse of green nanotechnology.

References

1. Bharadwaj, K.K., Rabha, B., Pati, S., Sarkar, T., Choudhury, B.K., Barman, A., ... Mohd Noor, N.H.: Green synthesis of gold nanoparticles using plant extracts as beneficial prospect for cancer theranostics. Molecules **26**(21), 6389 (2021).https://doi.org/10.3390/molecules26216389
2. Rónavári, A., Igaz, N., Adamecz, D.I., Szerencsés, B., Molnar, C., Kónya, Z., Pfeiffer, I., Kiricsi, M.: Green silver and gold nanoparticles: biological synthesis approaches and potentials for biomedical applications. Molecules **26**(4), 844 (2021). https://doi.org/10.3390/molecules26040844
3. Sarfraz, N., Khan, I.: Plasmonic gold nanoparticles (AuNPs): properties, synthesis and their advanced energy, environmental and biomedical applications. Chem. Asian J. (2021). https://doi.org/10.1002/asia.202001202.
4. Santhosh, P.B., Genova, J., Chamati, H.: Green synthesis of gold nanoparticles: an eco-friendly approach. Chemistry (2022).https://doi.org/10.3390/chemistry4020026
5. Zuhrotun, A., Oktaviani, D.J., Hasanah, A.N.: Biosynthesis of gold and silver nanoparticles using phytochemical compounds. Molecules (2023).https://doi.org/10.3390/molecules28073240
6. Ramli, M., Sujoko, I., Adhha, N., Annas, D., Nikmatullah, M., Ariyanta, H.A., ... Khairurrijal, K.: Influence of various concentrations of chloroauric acid on the fabrication of gold nanoparticles: Green synthesis using Elaeis guineensis Jacq leaf extract and characterizations. Results in Surfaces and Interfaces **17**, 100317 (2024). https://doi.org/10.1016/j.rsurfi.2024.100317
7. Wylie, M.R., Merrell, D.S.: The antimicrobial potential of the neem tree Azadirachta indica. Front. Pharmacol. (2022). https://doi.org/10.3389/fphar.2022.891535
8. Kumar, R., Saha, P., Lokare, P., Datta, K., Selvakumar, P., Chourasia, A.: A systemic review of Ocimum sanctum (Tulsi): Morphological characteristics, phytoconstituents and therapeutic applications. Int. J. Res. Appl. Sci. Biotechnol. **9**(2), 221–226 (2022). https://doi.org/10.31033/ijrasb.9.2.15.
9. Khan, R.U., Naz, S., De Marzo, D., Dimuccio, M.M., Bozzo, G., Tufarelli, V., ... Ragni, M.: Aloe vera: a sustainable green alternative to exclude antibiotics in modern poultry production. Antibiotics **12**(1), 44 (2022). https://doi.org/10.3390/antibiotics12010044
10. Moreira, P., Sousa, F.J., Matos, P., Brites, G.S., Gonçalves, M.J., Cavaleiro, C., ... Pereira, C.F.: Chemical composition and effect against skin alterations of bioactive extracts obtained by the hydrodistillation of Eucalyptus globulus leaves. Pharmaceutics **14**(3), 561 (2022).https://doi.org/10.3390/pharmaceutics14030561
11. Du, K., Zhang, Z., Jing, D., Wang, Y., Li, X., Meng, D.: Diterpene glycosides, acetophenone glycosides and tannins from polar extracts of the root of Euphorbia fischeriana with cytotoxicity

and antibacterial activities. Phytochemistry (2022). https://doi.org/10.1016/j.phytochem.2022.113382
12. Siddiqui, S.A., Singh, S., Nayik, G.A.: Bioactive compounds from pomegranate peels-Biological properties, structure–function relationships, health benefits and food applications–a comprehensive .... J. Funct. Foods (2024).https://doi.org/10.1016/j.jff.2024.106132
13. Dadoosh, R.M.: Evaluation of phytochemical, total phenolic and antioxidant activity of carica papaya seed for its use in biosynthesis of gold nanoparticles. Egypt. J. Chem. (2021). https://doi.org/10.21608/EJCHEM.2021.63985.3371
14. Ahmed, S.F., Mofijur, M., Rafa, N., Chowdhury, A.T., Chowdhury, S., Nahrin, M., ... Ong, H.C.: Green approaches in synthesising nanomaterials for environmental nanobioremediation: technological advancements, applications, benefits and challenges. Environ. Res. **204**, 111967 (2022). https://doi.org/10.1016/j.envres.2021.111967
15. Harish, V., Ansari, M.M., Tewari, D., Yadav, A.B., Sharma, N., Bawarig, S., ... Barhoum, A.: Cutting-edge advances in tailoring size, shape, and functionality of nanoparticles and nanostructures: a review. J. Taiwan Inst. Chem. Eng. **149**, 105010 (2023). https://doi.org/10.1016/j.jtice.2023.105010
16. Azad, A., Zafar, H., Raza, F., Sulaiman, M.: Factors influencing the green synthesis of metallic nanoparticles using plant extracts: a comprehensive review. Pharmaceutical Fronts (2023).https://doi.org/10.1055/s-0043-1774289
17. Ahmad, T., Iqbal, J., Bustam, M.A., Irfan, M., Asghar, H.M.A.: A critical review on phytosynthesis of gold nanoparticles: issues, challenges and future perspectives. J. Clean. Prod. **309**, 127460 (2021). https://doi.org/10.1016/j.jclepro.2021.127460
18. Santhosh, P.B., Genova, J., Chamati, H.: Green synthesis of gold nanoparticles: an eco-friendly approach. Chemistry (2022). https://doi.org/10.3390/chemistry4020026
19. Saini, N., Anmol, A., Kumar, S., Bakshi, M., Dhiman, Z.: Exploring phenolic compounds as natural stress alleviators in plants-a comprehensive review. Physiol. Molec. Plant Pathol., 102383 (2024).https://doi.org/10.1016/j.pmpp.2024.102383
20. Raspor, M., Motyka, V., Kaleri, A.R., Ninković, S., Tubić, L., Cingel, A., Ćosić, T.: Integrating the roles for cytokinin and auxin in de novo shoot organogenesis: from hormone uptake to signaling outputs. Int. J. Mol. Sci. **22**(16), 8554 (2021). https://doi.org/10.3390/ijms22168554
21. Motolinía-Alcántara, E.A., Castillo-Araiza, C.O., Rodríguez-Monroy, M., Román-Guerrero, A., Cruz-Sosa, F.: Engineering considerations to produce bioactive compounds from plant cell suspension culture in bioreactors. Plants **10**(12), 2762 (2021). https://doi.org/10.3390/plants10122762
22. Manjula, N.G., Tajunnisa, Mamani, V., Meghana, C.A., Mayegowda, S.B.: Fungal-based synthesis to generate nanoparticles for nanobioremediation. In: Green Nanoremediation: Sustainable Management of Environmental Pollution, pp. 83–108. Springer International Publishing, Cham (2023). https://doi.org/10.1007/978-3-031-30558-0
23. Rabiee, N., Ahmadi, S., Akhavan, O., Luque, R.: Silver and gold nanoparticles for antimicrobial purposes against multi-drug resistance bacteria. Materials (2022).https://doi.org/10.3390/ma15051799
24. Dhanker, R., Hussain, T., Tyagi, P., Singh, K.J., Kamble, S.S.: The emerging trend of bioengineering approaches for microbial nanomaterial synthesis and its applications. Front. Microbiol. **12**, 638003 (2021). https://doi.org/10.3389/fmicb.2021.638003

Chapter 5
Green Synthesis Using Microorganisms

Abstract A sustainable and environmentally friendly approach to nanotechnology is the green synthesis of gold nanoparticles (AuNPs) using microorganisms. The intracellular and extracellular reduction of gold ions (Au^{3+}) into stable metallic gold (Au^0) is achieved by bacteria, fungi, and marine microbes. Bacterial strains such as *Bacillus subtilis* and *Shewanella oneidensis* use enzymatic reduction to generate highly stable nanoparticles, which find applications in biomedical and environmental domains. Fungi such as *Aspergillus niger* and *Fusarium oxysporum* produce extracellular enzymes that facilitate the formation of nanoparticles with enhanced stability and biocompatibility. Marine organisms, adaptable to extreme conditons, including *Marinobacter hydrocarbonoclasticus*, have been employed for large-scale, cost-effective production. This method does not involve the use of toxic chemicals in the process, thereby making it safe for use in medicine, catalysis, and environmental applications. Despite the challenges of purification and standardization, microbial synthesis can be considered a scalable and sustainable alternative to chemical methods. This chapter will highlight the mechanisms, advantages, and industrial potential of microorganism-mediated AuNP synthesis and how it can be used to advance green nanotechnology as well as sustainable materials science.

5.1 Bacterial Synthesis of AuNPs

Bacteria-mediated production of gold nanoparticles (AuNPs) is a highly potent and environmentally friendly approach for the biological production of nanoparticles, in which microbes employ specific intracellular and extracellular mechanisms for the transformation of gold ions (Au^{3+}) to metallic gold (Au^0). The process does not require toxic chemicals, and may thus be suitable for biocompatible and scalable applications particularly in the biomedical and environmental fields [28]. In this biogenic reduction process, *Shewanella, Bacillus subtilis* and other microorganisms use reductase enzymes, proteins, and polysaccharides to reduce Au^{3+} to Au^0, initiate nanoparticles, and stabilize them thereby regulating their size and morphology [29].

A. C. Alguno et al., *Gold Nanoparticles*, Engineering Materials,
https://doi.org/10.1007/978-981-96-6565-5_5

Extracellular synthesis, as illustrated in Fig. 5.1a, is mediated through the secretion of the reductase enzyme that interact with gold ions to generate clustered nanoclusters. On the other hand, Fig. 5.1b depicts intracellular synthesis, wherein Au^{3+} ions are internally introduced into bacterial cells and reduced in the cytoplasm. Both mechanisms results in the successful formation of nanoparticles, although they have distinct reaction kinetics and microenvironments.

The synthesis starts with a gold salt solution (usually $NaAuCl_4$), which progresses from pale-yellow, to colorless (after a reduction of Au^{3+}), and may involve a blue-gray preliminary product of bluish-gray nanowires, preceding the spherical nanoparticles, although nanowires are not always evident [30]. Due to factors such as pH, temperature, and gold ion concentration, this synthesis can be finely tuned, demonstrating sustainable innovation in accordance with principles of green chemistry [31].

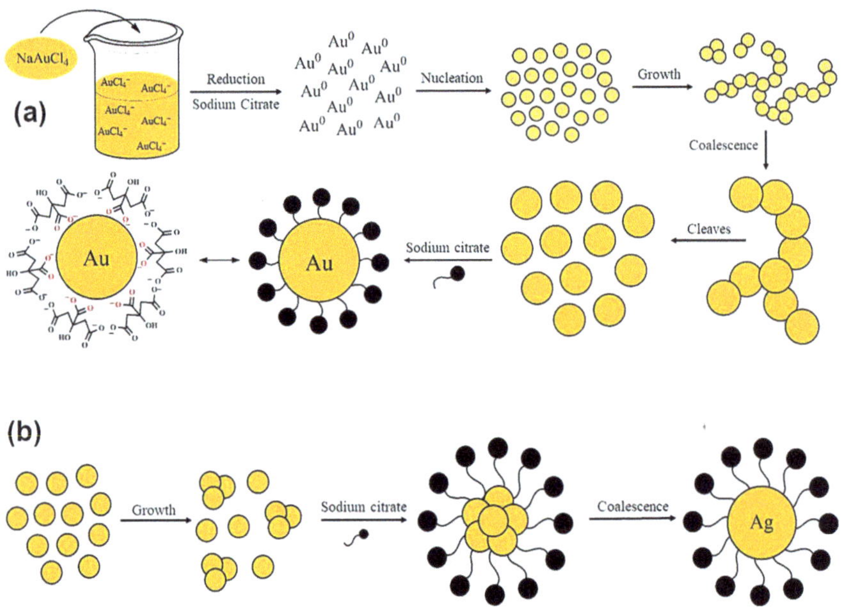

Fig. 5.1 Mechanisms of AuNPs synthesis in the Turkevich method. **a** Formation of intermediaries, and **b** non-aggregate intermediates, leading to nanoclusters, nanowire networks, and eventual spherical particles through rapid reduction and coalescence processes. Adapted from Oliveira et al. (2023), Fig. 5.1, https://doi.org/10.3390/analytica4020020. Licensed under CC BY 4.0 (https://creativecommons.org/licenses/by/4.0/)

5.1.1 *Mechanisms and Case Studies:* **Bacillus, Shewanella,** *and* **Marinobacter** *Species*

Bacterial synthesis of AuNPs is a good example of how microorganisms can be used as environmentally friendly agents in the production of nanotechnology. The following are some of the isolates that have been identified to produce AuNPs, and the role they play in different applications include: *Bacillus subtilis*, *Shewanella oneidensis* and *Marinobacter hydrocarbonoclasticus*. *Bacillus subtilis* uses reductase enzymes to reduce Au^{3+} ions to Au^0, and uses extracellular polysaccharides to stabilize and control the size of the nanoparticles for better catalysis and electronics [1]. Conversely, *Shewanella oneidensis* is a metal reducing and oxidizing bacterium that uses cytochromes to transfer electrons in the formation of nanoparticles of various sizes and shapes, which are useful in biomedical applications, such as drug delivery systems [2]. In addition, *Marinobacter hydrocarbonoclasticus* exhibits a high capacity to synthesize nanoparticles of gold in saline conditions with the help of extracellular polymeric substances as reducing and capping agents, and this is important in environmental protection [3].

This comparative study elucidates the extraordinary adaptive responses of these bacterial species towards diverse environmental perturbations, in addition to tailoring the particles for desired applications. The main conclusion from the studies compiled here is that AuNPs synthesized biologically by bacteria serve as an eco-friendly, cost-effective, and scalable method compared to chemical processes, and represent an important step in green nanotechnology. These insights, in turn, provide directions for the sustainable production of AuNP, and importantly, link biological creativity with industrial applications.

5.2 Fungal Synthesis of AuNPs

Using fungi in the reduction of Au^{3+} ions is a step forward in nanomaterial generation due to the ability of fungi to secrete extracellular enzymes or extracellular metabolites, which reduce and stabilize AuNPs, thereby facilitating their fungal-mediated synthesis. The species which are becoming preferable during this process are *Aspergillus niger* and *Fusarium oxysporum,* which excel throughout this process due to the production of high biomass and proteins like nitrate reductase. This bioreduction reduces gold ions into Au^0 nanoparticles in favorable conditions, to yield nanoparticles of well-defined morphology and size. Besides enabling the process to be up-scaled, it is also less toxic and more environmentally friendly than traditional methods, resulting in fungal synthesis becoming one of the most preferred methods for the generation of biocompatible nanoparticles. These characteristics are especially advantageous for medical imaging and targeted drug delivery applications, which require non-toxic, efficient nanomaterials. Therefore, fungal-mediated synthesis promises the development of new biomedical technologies.

5.2.1 Role of Extracellular Enzymes

Extracellular enzymes are very important in the green synthesis of nanoparticles, especially AuNPs, because they help in the formation of the nanoparticles from metal ions into elemental gold nanoparticles. This mechanism is demonstrated by fungi such as *Fusarium oxysporum,* which secretes reductases that reduce Au^{3+} to Au^{+}, and does so at mild environmental conditions, whereas extracellular polysaccharides act as biological stabilizers in this biotransformation process [4]. These enzymes are a regulating factor in the process of nucleation, which leads to the development of monodisperse nanoparticles of utility in biomedical and environmental science [5].

Fungi are known to secrete a large number of extracellular enzymes and other biomolecules that make them well suited for this task. Nitrate reductase and laccase are crucial enzymes in this regard, as they catalyze electron transfer reactions that are essential for the reduction of metal ions. In addition, several proteins and metabolites help in the stabilization of the nanoparticles, and control their shape [6]. Therefore, fungal systems can be recommended as superior for green synthesis, being environmentally friendly and scalable, in comparison with conventional chemical methods [7]. Enzymatic reduction and biomolecular capping are shown to interact synergistically towards the development of sustainable nanotechnology, and the production of biocompatible materials with a wide range of industrial and biomedical uses.

5.2.2 Examples: **Penicillium, Aspergillus,** *and* **Macrophomina** *Species*

Fungal species such as *Penicillium* and *Aspergillus* have been explored in the synthesis of AuNPs [8], which has received a considerable amount of attention in the scientific community because of their tendency to produce extracellular enzymes and metabolites in large quantities. For example, *Penicillium spp.* produce nitrate reductase enzymes that efficiently reduce Au^{3+} ions, resulting in homogeneous nanoparticle shapes and sizes [9]. These well-characterized nanoparticles are used in cancer therapy and biosensing, indicating clinical relevance.

Conversely, AuNP synthesis by *Aspergillus niger* is enzymatic and protein mediated, as stated by [10] and the nanoparticles produced have great stability, and also have optimum optical properties. Such features make it highly useful for photothermal therapy and diagnostic purposes. It also indicates that AuNPs can be synthesized in the absence of cells via the action of an enzyme, such as *Macrophomina phaseolina,* and the AuNPs thus synthesized appear to be potent antimicrobial agents and effective catalysts [4]. Moreover, *Macrophomina-derived* AuNPs can be applied for the degradation of organic toxins present in wastewater treatment, which becomes an exemplification of such oxidation in real life.

These cases, therefore, illustrate the great possibility of using fungi in the synthesis of nanoparticles with different features in various applications [11]. As such, this biotechnological process is eco-friendly, and biocompatible, and therefore offers a good alternative to conventional chemical synthesis methods that are more consistent with the principles of green chemistry and sustainability in nanotechnology.

5.3 Green Synthesis Using Marine Microorganisms

Marine microorganisms have proven to be significant when it comes to the green route for the synthesis of AuNPs, establishing their role as key agents in sustainable nanotechnology. This is exemplied by *Aureobasidium pullulans* strains, as well as *Bacillus aquimaris*, and *Aspergillus sydowii,* all of which adapt well to extreme saline conditions and can promote the reduction of gold ions (Au^{3+}) into stable AuNPs. The change is mediated by a variety of extracellular biomolecules, including proteins, polysaccharides, and secondary metabolites, which effectively act as dual-purpose reducing and capping agents. These biomolecules provide stability to the nanoparticles, and regulate their size and shape [12, 32].

Industrial enzymes, for instance, used in the marine sector, are significant for their environmental advantages and compatibility. Using saline-adapted enzymes, these organisms produce AuNPs at high concentrations, resulting in biocompatible AuNPs that can be employed in drug delivery systems, catalysis, and environmental remediation. Moreover, the mild conditions used for the synthesis of nanoparticles also help in eliminating the use of hazardous reagents, in accordance with principles of green chemistry [13].

Marine biotechnology and nanoparticle synthesis represent an integration of environmental sustainability and cutting-edge science. With this, research in this area continues to advance, there is potential to redefine conventional nanoparticle synthesis, forming the core of scalable, nature-inspired methods that are not only eco-friendly, but also versatile, and applicable to broadening spectrum of industrial applications, from medicine to electronics, thus placing marine microorganisms at the forefront of stimulating the development of sustainable nanoscience.

5.3.1 Potential of Marine Bacteria and Fungi

The biochemical diversity and the specific enzymatic activities of marine bacteria and fungi are exploited for the biosynthesis of AuNPs. Moreover, gram-negative marine bacteria such as *Marinobacter* and *Shewanella* use extracellular polymers in association with reductase enzymes for the reduction and stabilization of gold ions, and yield stable nanoparticles with tunable properties [14, 15]. This demonstrates their comparative advantage in biomedical and catalytic applications.

Meanwhile, marine fungi like *Penicillium chrysogenum* and *Fusarium solani* are used to synthesize AuNPs [16, 17]. Several fungi release enzymes that rapidly reduce and stabilize gold ions, such as nitrate reductase. Over the course of evolution under extreme marine conditions, these fungi have adapted to such stressors, which can contribute to nanoparticles with unique morphology. Gradually, these biogenic AuNPs generated from these microorganisms are being utilized in many applications, including targeted drug delivery systems, biosensing devices, and environmental remediation.

This chapter presents case studies to support the idea that nanoparticles produced by marine microbes usually have improved catalytic and antimicrobial activity, due to natural capping agents from microbial exudates [10, 18]. This particular aspect is significant, as it highlights the potential of these biomaterials for sustainable nanotechnology development in the areas of energy and healthcare. Therefore, using marine microorganisms, it is possible to develop novel solutions that would be environmentally friendly, at the same time leverage nanotechnology for significant social change.

5.3.2 Advantages of Marine-Based Synthesis

With this new method, we investigate an eco-friendly approach that not only generates more sustainable nano-beta galactosidase, but also, due to its marine origin, provides high efficiency [19]. Marine microorganisms are engineered to grow in salty mediums which eliminates the need for strict sterile deals significantly lowering up the production costs. These organisms synthetize polysaccharides and proteins, which act as efficient capping agents, contributing to the stability and biocompatibility of the produced nanoparticles [20].

Marine derived AuNPs have been found to have different physicochemical properties compared to conventionally synthesized nanoparticles as a result of the extremophilic microorganisms that produce them. For example, nanoparticles produced from *Marinobacter hydrocarbonoclasticus* display high catalytic activity in the degradation of organic pollutants, and therefore significant potential in environmental technologies [21]. The process of biosynthesis can be regarded as one of the most environmentally friendly approaches, as it does not involve the use of toxic substances, and does not lead to the formation of toxic wastes.

In addition, the scalability of this method of synthesis is advantegeous; marine microorganisms grow in normal conditions, and produce high yields, which is important for large-scale production [22]. The biological molecules that are involved in the nanoparticle synthesis also improve the optical and plasmonic characteristics of the nanoparticles, thus making them suitable for applications in biosensing and photothermal therapy.

5.4 Optimization and Control of Nanoparticle Properties

Similarly, AuNP synthesis is optimized based on key parameters: pH, temperature, and the concentration of ions. One such organism includes marine microorganisms, which possess unique enzyme pathways that enable them to exert precise control on nanoparticle sizes and stability. Researchers were able to control every aspect of the nanoparticle functionality by fine-tuning the following aspects of the culture conditions, such as salinity and nutrient availability throughout the growth phases, subsequently customizing nanoparticle characteristics for targeted applications. For example, *S. oneidensis* and members of the genus *Marinobacter* have been shown to form relatively spherical AuNPs in neutral conditions, and triangular structures in acidic conditions, respectively. This remarkable tunability not only improves the functional diversity of biogenic AuNPs, but also makes them valuable as templates for catalytic, imaging, and delivery applications.

5.4.1 Influence of Culture Conditions, Growth Media, and Incubation Time

It is crucial to optimize culture conditions, growth media, and incubation times in the biosynthesis of AuNPs by microorganisms. The temperature, pH, and salinity are critical parameters that affect the reduction of gold ions and the quality of the nanoparticles. However, *Marinobacter hydrocarbonoclasticus* does best when the temperature is 30–37 °C, the pH is 7.0–8.5, and the salt is 3–5% NaCl. This helps maintain the nanoparticles stable, and controls their size and shape [23, 32].

However, the type of growth medium is also particularly important. The addition of carbon and nitrogen sources, such as peptone and yeast extract, increases the enzyme concentration, and, therefore, the rate of gold-ion reduction. *S. oneidensis*, for instance, produces spherical nanoparticles of 10–20 nm in diameter when grown in acetate containing medium [24].

Furthermore, the incubation time is the most important factor that determines the size and shape of the nanoparticles; short times (12–24 h) lead to the formation of small, monodispersed nanoparticles, whereas long times (48 h or above) result in large nanoparticles with different shapes, such as rod and prism [33]. The precise control of synthesis parameters is not only sessential for tuning AuNP properties for their potential use in biomedicine and environmental analysis, but also for obtaining high NP yields, good biocompatibility, and reproducibility, which are factors that are critical for the development of industrial-scale production.

5.5 Pros and Cons of Microbial Synthesis

The production of AuNPs by microbial methods is a novel and environmentally friendly approach that is preferable to the conventional chemical methods because it is cheaper, and biocompatible. This method not only scales up very efficiently, but also has a low environmental impact through the use of natural microbial capping agents, which increase the stability and functionality of the NPs. However, there are some limitations, such as the risk of contamination, and the difficulty in the purification of biological residues from the final product. Moreover, the current work has revealed that the size and stability of the nanoparticles vary with the type of microbial strain, and synthesis conditions. These hurdles can only be overcome through the application of novel bioprocessing strategies, to obtain reproducibility, and enhance the quality of the product for sustainable production of nanotechnology.

5.5.1 Advantages: High Yield and Eco-Friendliness

Microbial production is the most effective approach to green nanotechnology, as it provides high yield, and environmentally friendly production. Marine bacteria are the leading example, as they can produce nanoparticles in large quantities with moderate temperature, which in turn reduces energy consumption. Using non-toxic biological systems as the reducing and stabilizing agents does not only avoid the generation of harmful by-products, but also follows the concepts of green chemistry, thus recommending this method as a less toxic approach for production processes [25]. This approach is also sustainable, and meets the needs of today's world for its use in industry and, at the same time, it is safe for the environment and people. In addition, the capping layers of the nanoparticles, which are biocompatible, and are supplied by the microbes, improve the functional properties of the nanoparticles significantly [26]. These are especially important in biomedical and catalytic systems where the material properties are critical. Therefore, microbial synthesis can satisfy the requirements of present scientific paradigms, and create a basis for future innovations that will be compatible with award-winning scientific discourse and application in sustainable technology [27].

5.5.2 Limitations: Contamination Risk and Purification Challenges

Microbial synthesis of AuNPs is plagued by contamination and purification issues due to the co-synthesis of extracellular macromolecules, such as proteins, polysaccharides, and secondary metabolites. The nanoparticle surface can change in size, shape, and functionality when biomolecules attach. These limitations are directly obstructive to the deployment of the nanoparticles in different areas, because of the changes they bring to the nanoparticles [12, 34]. Contamination occurs from metabolites that are produced during microbial growth, and is especially a problem when conditions are less than ideal.

For instance, *Bacillus* species are effective in reducing Au^{3+} to Au^0, but produce proteins that can modify the nucleation and stabilization of the nanoparticles randomly. This results in complicated and time-consuming purification methods, such as centrifugation and dialysis [32]. Similarly, the nanoparticles generated by *Shewanella oneidensis* are heterogeneous in size and dispersion, because of the complex and variable electron transfer systems that are used in their production, thus limiting their application in biomedical fields [33]. A recent study by Kusuma et al. (2023) [35] showed that unpurified AuNPs generated by *Marinobacter* species had biomolecular impurities that altered the optical features of the nanoparticles, thus rendering them less efficient in colorimetric detection assays.

To overcome the contamination risks, advanced purification techniques have been suggested, for instance, using ultrafiltration in combination with chromatographic methods; however, such methods may increase the complexity and costs of the process [36]. Standardized microbial synthesis protocols, and genetic engineering to prevent the formation of unwanted by-products, are necessary to solve these problems. As such, the formation of robust strategies will enable the production of nanoparticles of predictable quality, which will in turn allow for the full exploitation of their advantages in environmentally friendly diagnostic and therapeutic purposes, and other industries.

References

1. Arnaouteli, S., Bamford, N.C., Stanley-Wall, N.R., Kovács, Á.T.: Bacillus subtilis biofilm formation and social interactions. Nat. Rev. Microbiol. **19**(9), 600–614 (2021). https://doi.org/10.1038/s41579-021-00540-9
2. Yin, Y., Liu, C., Zhao, G., Chen, Y.: Versatile mechanisms and enhanced strategies of pollutants removal mediated by Shewanella oneidensis: a review. J. Hazardous Mater. (2022). https://doi.org/10.1016/j.jhazmat.2022.129703
3. Gaur, V.K., Sharma, P., Gupta, S., Varjani, S., Srivastava, J.K., Wong, J.W., Ngo, H.H.: Opportunities and challenges in omics approaches for biosurfactant production and feasibility of site remediation: strategies and advancements. Environ. Technol. Innov. **25**, 102132 (2022). https://doi.org/10.1016/j.eti.2021.102132
4. Saravanan, A., Kumar, P.S., Karishma, S., Vo, D.V.N., Jeevanantham, S., Yaashikaa, P.R., George, C.S.: A review on biosynthesis of metal nanoparticles and its environmental applications. Chemosphere **264**, 128580 (2021). https://doi.org/10.1016/j.chemosphere.2020.128580
5. He, M.Q., Ai, Y., Hu, W., Guan, L., Ding, M., Liang, Q.: Recent advances of seed-mediated growth of metal nanoparticles: from growth to applications. Adv. Mater. **35**(46), 2211915 (2023). https://doi.org/10.1002/adma.202211915
6. Chugh, D., Viswamalya, V.S., Das, B.: Green synthesis of silver nanoparticles with algae and the importance of capping agents in the process. J. Genetic Eng. Biotechnol. **19**(1), 126 (2021). https://doi.org/10.1186/s43141-021-00228-w

7. Olaiya, N.G., Oyekanmi, A.A., Hanafiah, M.M., Olugbade, T.O., Adeyeri, M.K., Olaiya, F.G.: Enzyme-assisted extraction of nanocellulose from textile waste: a review on production technique and applications. Bioresource Technol. Reports **19**, 101183 (2022). https://doi.org/10.1016/j.biteb.2022.101183
8. Xu, F., Li, Y., Zhao, X., Liu, G., Pang, B., Liao, N.,... Shi, J.: Diversity of fungus-mediated synthesis of gold nanoparticles: properties, mechanisms, challenges, and solving methods. Critical Rev. Biotechnol. **44**(5), 924–940 (2024). https://doi.org/10.1080/07388551.2023.2225131
9. Zhang, R., Yin, J., Sui, Z., Han, L., Li, Y., Huang, J.: Biocontrol of antifungal volatiles produced by Ceriporia lacerate HG2011 against citrus fruit rot incited by Penicillium spp. Postharvest Biol. Technol. **194**, 112094 (2022). https://doi.org/10.1016/j.postharvbio.2022.112094
10. Koul, B., Poonia, A.K., Yadav, D., Jin, J.O.: Microbe-mediated biosynthesis of nanoparticles: applications and future prospects. Biomolecules (2021).https://doi.org/10.3390/biom11060886
11. Kumar, P., Thakur, N., Kumar, K., Kumar, S., Dutt, A., Thakur, V.K.,... Thakur, N.: Catalyzing innovation: exploring iron oxide nanoparticles-Origins, advancements, and future application horizons. Coordination Chem. Rev. **507**, 215750 (2024). https://doi.org/10.1016/j.ccr.2024.215750
12. Santhosh, P.B., Genova, J., Chamati, H.: Green synthesis of gold nanoparticles: an eco-friendly approach. Chemistry (2022).https://doi.org/10.3390/chemistry4020026
13. Pereira, L., Cotas, J.: Seaweed: a sustainable solution for greening drug manufacturing in the pursuit of sustainable healthcare. Exploration Drug Sci. (2024). https://doi.org/10.37349/eds.2024.00036.
14. Chakraborty, N., Jha, D., Roy, I., Kumar, P., Gaurav, S.S., Marimuthu, K.,... Gautam, H.K.: Nanobiotics against antimicrobial resistance: harnessing the power of nanoscale materials and technologies. J. Nanobiotechnology **20**(1), 375 (2022).https://doi.org/10.1186/s12951-022-01573-9
15. Ghosh, A., Sah, D., Chakraborty, M., Rai, J.P.N.: Mechanism and application of bacterial exopolysaccharides: an advanced approach for sustainable heavy metal abolition from soil. Carbohydrate Res. (2024).https://doi.org/10.1016/j.carres.2024.109247
16. Ahmed, F.K., Kalia, A., Ahmad, A., Alghuthaymi, M.A., Abd-Elsalam, K.A.: Marine fungi and yeast: a green approach for production of bionanoparticles. In: Fungal Cell Factories for Sustainable Nanomaterials Productions and Agricultural Applications, pp. 337–360. Elsevier (2023)
17. Bhandari, Y., Varma, S., Sawant, A., Beemagani, S., Jaiswal, N., Chaudhari, B.P., Vamkudoth, K.R.: Biosynthesis of gold nanoparticles by Penicillium rubens and catalytic detoxification of ochratoxin A and organic dye pollutants. Int. Microbiol. **26**(4), 765–780 (2023). https://doi.org/10.1007/s10123-023-00341-5
18. Shunmugam, R., Balusamy, S.R., Kumar, V., Menon, S., Lakshmi, T., Perumalsamy, H.: Biosynthesis of gold nanoparticles using marine microbe (Vibrio alginolyticus) and its anticancer and antioxidant analysis. J. King Saud Univ. Sci. **33**(1), 101260 (2021). https://doi.org/10.1016/j.jksus.2020.101260
19. Rizki, I.N., Klaypradit, W.: Utilization of marine organisms for the green synthesis of silver and gold nanoparticles and their applications: a review. Sustain. Chem. Pharmacy (2023). https://doi.org/10.1016/j.scp.2022.100888
20. Yosri, N., Khalifa, S.A., Guo, Z., Xu, B., Zou, X., El-Seedi, H.R.: Marine organisms: Pioneer natural sources of polysaccharides/proteins for green synthesis of nanoparticles and their potential applications. Int. J. Biol. Macromol. **193**, 1767–1798 (2021). https://doi.org/10.1016/j.ijbiomac.2021.10.229
21. Liu, S., Pan, L., Chen, J., Wang, Z., Li, Z., Gao, C., Yang, H.: Mechanisms and applications of microbial synthesis of metal nanoparticles in agri-sectors. Environ. Sci. Nano **11**(7), 2803–2830 (2024). https://doi.org/10.1039/D4EN00174E
22. Ahmed, S.F., Mofijur, M., Rafa, N., Chowdhury, A.T., Chowdhury, S., Nahrin, M.,... Ong, H.C.: Green approaches in synthesising nanomaterials for environmental nanobioremediation: technological advancements, applications, benefits and challenges. Environ. Res. **204**, 111967 (2022). https://doi.org/10.1016/j.envres.2021.111967

23. Singh, H., Desimone, M.F., Pandya, S., Jasani, S., George, N., Adnan, M.,... Alderhami, S.A.: Revisiting the green synthesis of nanoparticles: uncovering influences of plant extracts as reducing agents for enhanced synthesis efficiency and its biomedical applications. Int. J. Nanomed., 4727–4750 (2023).https://doi.org/10.2147/IJN.S419369
24. Coviello, V., Forrer, D., Amendola, V.: Recent developments in plasmonic alloy nanoparticles: synthesis, modelling, properties and applications. ChemPhysChem (2022).https://doi.org/10.1002/cphc.202200136
25. Jain, K., Takuli, A., Gupta, T.K., Gupta, D.: Rethinking nanoparticle synthesis: a sustainable approach vs. traditional methods. Chem. Asian J., e202400701 (2024). https://doi.org/10.1002/asia.202400701
26. Namasivayam, S.K.R., John, A., RS, A.B., Kavisri, M., Moovendhan, M.: Biocompatible formulation of cationic antimicrobial peptide Polylysine (PL) through nanotechnology principles and its potential role in food preservation—a review. Int. J. Biol. Macromolecules **222**, 1734–1746 (2022).https://doi.org/10.1016/j.ijbiomac.2022.09.238
27. Pattoo, T.A.: Flora to nano: sustainable synthesis of nanoparticles via plant-mediated green chemistry. Plant Sci. Arch. (2023). https://doi.org/10.51470/PSA.2023.8.1.12
28. Sandeep, T., Kumar N., Naga, M., Rao, R., Rawat, H., Surekha, R., Manu, S., Veera, S., Annadanam, V., Sesha, S.: Galactopolymer architectures/functionalized graphene oxide nanocomposites for antimicrobial applications. Journal of Polymer Research. **28**(6) (2021). https://doi.org/10.1007/s10965-021-02528-8
29. Kannan, B., Narayanan, N., S.: Biological synthesis of metal nanoparticles by microbes Advances in Colloid and Interface. Science. **156**(1-2) 1–13 (2010). https://doi.org/10.1016/j.cis.2010.02.001
30. Subramaniyam, S., Kiran, D., Madhuri, B., Tarun, M., Adele, S., Biswanath, M., Priyadip, D.: Self-assembled dipeptide based fluorescent nanoparticles as a platform for developing cellular imaging probes and targeted drug delivery. chaperones Nanoscale Advances. **4**(6) 1694–1706 (2022). https://doi.org/10.1039/D1NA00885D
31. Ahmed, B., Jaison, J., Amit, R., Pieter, S., Yaman, B., Alain, D., Michael, K., Danquah, M., B.: Plant celluloses hemicelluloses lignins and volatile oils for the synthesis of nanoparticles and nanostructured materials. Nanoscale. **12**(45) 22845–22890 (2020). https://doi.org/10.1039/D0NR04795C
32. Rai, M., & Posten, C. (Eds.). Green biosynthesis of nanoparticles: Mechanisms and applications. CAB International. (2014). https://doi.org/10.1079/9781780642239.0000
33. Myalowenkosi, S., Sebastien, B., Mikhael, B., Jean-Marc, J., Krishna, B.: A Review of Gold and Silver Nanoparticle-Based Colorimetric Sensing. Assays Advanced Engineering Materials. **19**(12) (2017). https://doi.org/10.1002/adem.v19.12
34. Andrea, R., Nóra, I., Dóra, I., Adamecz, B., Szerencsés, C., Molnar, Z., Kónya, I., Pfeiffer, M., K.: Green Silver and Gold Nanoparticles: Biological Synthesis Approaches and Potentials for Biomedical. Applications Molecules. **26**(4) 844 (2021). https://doi.org/10.3390/molecules26040844
35. Sri, A., Fitri, K., Jacko, A., Harmonis, R., Pratiwi, A., Nur, H.: Gold Nanoparticle-Based Colorimetric Sensors: Properties and Application in Detection of Heavy Metals and Biological. Molecules Sensors. **23**(19) 8172 (2023). https://doi.org/10.3390/s23198172
36. Daniel L., Fedlheim C., Foss, A.: (Eds.). Metal nanoparticles: Synthesis, characterization, and applications. CRC Press. (2001). https://doi.org/10.1201/9780367800475

Chapter 6
Properties of Gold Nanoparticles and Their Functionalization

Abstract Gold nanoparticles (AuNPs) are nanoscale particles of gold with distinct physical, optical, and chemical properties that are crucial for nanotechnology. Localized surface plasmon resonance (LSPR) is a phenomenon of collective electron oscillation in noble metals, and offers single-molecule detection, hence, they find their application in biosensing, imaging, and photothermal therapy. The optical behavior is a function of size, shape, and the surrounding medium of the AuNPs, which can be used to tune the plasmonic response for specific applications. Chemical functionalization is performed to improve AuNP stability, selectivity, and reactivity. Surface enhancements, such as ligand replacement and bioattachment, are applied to improve biocompatibility and targeting for drug delivery, catalysis, and biosensing. The functionalized AuNPs have been applied in colorimetric sensing, cancer therapy, and environmental monitoring. However, the toxicity of the nanoparticles is an issue that depends on size, surface charge, and coatings of the particles. These risk factors are reduced through AuNP surface engineering, to enhance the use of AuNPs in medicine and industry. In this chapter, basic principles and functionalization strategies of AuNPs, and their major areas of application are reviewed with an emphasis on the role of these nanoparticles in nanomedicine, diagnostics, and sustainable emission.

6.1 Physical and Optical Properties

AuNPs have distinct physical, and optical characteristics. This is attributed to their nanoscale size, and their excellent electrical conductivity. Localized surface plasmon resonance (LSPR) is one of the crucial properties that involves the resonance of conduction electrons as a whole in response to certain wavelengths of light. Depending on the size, shape, and degree of aggregation of the nanoparticles, this phenomenon results in a high absorption band in the visible spectrum, and colors visible in the solution could be red, purple or blue [25, 26]. LSPR intensity and position are also greatly affected by the nanoparticle surroundings; for instance, spherical AuNPs have a sharp peak at about 520 nm, but elongated shapes, like nanorods, can

A. C. Alguno et al., *Gold Nanoparticles*, Engineering Materials,
https://doi.org/10.1007/978-981-96-6565-5_6

push this resonance well above 700 nm. These optical changes enrich the application of AuNPs in biosensors, to detect small changes in the environment through color changes [27, 28]. Also, they have high molar extinction coefficients (up to 109 M^{-1} cm^{-1}), and are photothermally stable and biocompatible, so AuNPs are used in biomedical imaging, photothermal therapy, and environmental monitoring. The specific optical properties of AuNPs are necessary for precise control in new developments in various scientific, and technological areas.

6.1.1 Size, Shape, and Surface Plasmon Resonance (SPR)

AuNPs are the most commonly used nanoparticles, because of their optical properties, which are a function of size, shape, and surface plasmonic resonance (SPR). SPR is the coherent oscillation of conduction electrons induced by electromagnetic radiation, which leads to strong absorption and scattering of light. This phenomenon is employed in a wide range of applications [1]. The SPR is typically centered at about 520 nm for spherical AuNPs. Fundamentally, small particles have large spectral shifts, but all are blue shifted due to their higher electron density, and resonance frequency, whereas larger particles are red shifted due to increasing resonance damping, and scattering [2, 3].

The SPR properties of AuNPs are also dependent on the geometric shape of the nanoparticles. Rod-shaped nanoparticles have two SPR bands, that correspond to their longitudinal, and transverse oscillations. The morphological flexibility of AuNPs enables them to be designed for photothermal therapies and biosensing devices, as stated in the reference [27].

Also, the dielectric environment of AuNPs is crucial and has a great impact on their optical performance. Alterations in the refractive index of this environment can lead to a shift in the SPR peak, and in for most cases, the increase in the refractive index results in red shift, which is advantageous for biosensing. Furthermore, the spatial arrangement of the neighboring particles is also reported to affect plasmon coupling, that results in aggregation related red shifts, and changes in the color of the solution. These principles are employed in colorimetric sensors, where the change in color of the AuNP solution is attributed to the aggregation of the nanoparticles upon the addition of the analyte, and therefore, can be used to rapidly detect biomolecules or contaminants [4].

The proper control of size, shape, and the environment of AuNPs is critically important for optimum SPR. This optimization also improves their functional application in imaging, diagnostics, and environmental monitoring, and thus justifies their classification as important nanoparticles in the field of nanotechnology research and application. In this manner, while optimizing the size, shape, and spacing of AuNPs, they are still contributing to the improvement of research in different scientific disciplines.

6.1.2 AuNP Role in Optical Applications and Bioimaging

Due to the particular SPR-induced optical properties of AuNPs, they are responsible for a wide range of optical applications. AuNPs also enhance contrast in advanced bioimaging techniques, such as optical coherence tomography (OCT), and multi-photon microscopy, through the modulation of light absorption and scattering by their surface [5]. Their extraordinarily high values of molar extinction coefficients allow them to be employed for the sensitive detection of cellular components, and biomolecules. Furthermore, when functionalized with specific ligands, such as antibodies, AuNPs enable selective binding to target biomarkers, resulting in high spatial resolution through imaging techniques.

Colorimetric sensing is another fascinating tool, operating on the simple concept of the visible colorimetric changes that result from analyte-induced aggregation [6]. Thiol- or peptide-functionalized AuNPs, for example, are versatile platforms and serve as markers for practical detection of a variety of substances, such as toxins, proteins, or nucleic acids, with changes even observable without the use of sophisticated tools [7]. Moreover, the generation of heat through SPR by AuNPs also drives photothermotherapy as a targeted technique for ablation of cancer cells. The benefit of their compatibility with near-infrared light is that they penetrate deep into the tissue to avoid collateral damage to healthy tissues. These diverse applications illustrate the contribution of AuNPs in facilitating the developments of optical technologies, and bioimaging.

6.2 Chemical and Surface Properties of AuNPs

AuNPs have distinct chemical properties, due to the presence of an inert gold core and the high reactivity of the surface. They exhibit high reactivity with organic, and inorganic molecules, through bonding via Au–S and Au–N bonds to functionalize the surface of the nanoparticles. Such functionalization is not only useful for selective targeting and colloidal stabilization, but also enhances the catalytic activity of the nanoparticles. The surface contains sites that are known to facilitate electron transfer, an important reaction for many applications. Furthermore, AuNPs show excellent biocompatibility and are therefore highly suitable for biomedical applications, such as drug delivery, and diagnostic imaging, and biosensing, as well as for environmental purposes, including pollution detection and remediation. The versatility of AuNPs, along with their intrinsic optical, electronic, and physicochemical properties, is thus demonstrated to hold promise for advances in science and technology.

6.2.1 *Reactivity, Surface Charge, and Functional Groups*

The intrinsic elevated surface free energy and abundance of active sites on AuNPs collectively contribute to their exceptional reactivity [8]. The extent of this electrostatic effect is critically determined by the surface charge, which is achieved through surface functionalization. When coated with negatively charged citrate ions [9], the AuNPs exhibit interparticle repulsion, thereby preventing aggregation of the particles via electrostatic interactions. The functional attributes can be significantly enhanced by replacing citrate ions with alternative ligands such as thiols or amines, thereby offering finer control over surface charge and interfacial behavior.

In addition, the introduction of specific functional groups through surface modifications significantly increases the functionality of AuNPs. The thiol-terminated molecules form stable covent linkages with gold surfaces, enabling the conjugation of biologically relevant macromolecules, including proteins and DNA [10]. This molecular alteration of the surface chemistry enables selective binding to target analytes, which is a requirement for advancing biosensing flatforms.

In addition to their extraordinary biosensing abilities, AuNPs are also important in catalysis. The catalytic activities of noble metal nanoparticles (NPs) are distinctly unique in that electrons are transferred through active sites for faciliting electron transfer, and reactivity is controlled prcisely by the particle size and surface chemistry of the nanoparticles. Typically, smaller AuNPs are more catalytically reactive because of their high surface area [11]. These nanoparticles are essential components in diagnostic, catalytic, and environmental monitoring applications, as well as core elements in nanotechnology research, all owing to the ability to control surface charge, reactivity, and surface functionalization for enhanced reactivity of AuNPs.

6.2.2 *Surface Modification Techniques for Enhanced Stability of AuNPs*

Due to the dependence in size, shape, optical properties, and reactive properties of AuNPs, they have found a significant place in the field of nanotechnology. The SPR is the collective oscillation of conduction electrons at the surface of the nanoparticle, induced by incident light, which results in a strong absorption peak and scattering signal. In the case of spherical AuNPs, these peaks typically occur in the range of approximately 520 nm [1]. Even small changes in particle size can lead to distinct SPR behavior, as smaller AuNPs show a blue-shift in resonance due to an increase in oscillation frequency, whereas larger AuNPs demonstrate a red -shift as a result of enhanced scattering, increased radiated damping, and multipolar interactions [8, 9].

Nanoparticle shape is also critical in influencing SPR behavior. For instance, rod-shaped AuNPs exhibit two distinct SPR peaks corresponding to transverse and longitudinal oscillations, and are used in photothermal therapy, plasmonic imaging, and advanced sensing techniques as tunable parameters [10]. However, the dielectric environment surrounding the AuNPs is also important in determining their optical properties; even a small change in the refractive index or the spatial distribution of the neighboring particles is sufficient to perturb the SPR peaks. This effect is especially observed during analyte-induced aggregation, which leads to plasmon coupling and the shift of the SPR peak to longer wavelength, typically accompanied by the appearance of a distinct colorimetric signal. Such behavior is utilized in biosensors, where aggregation is characterized by the significant shift in color from red to blue [4].

Surface functionalization is crucial for the stabilization of AuNPs, as well as for improving their chemical versatility and functional properties. In this context, ligand exchange means replacing the initial citrate capping with thiol-based ligands (e.g., methoxy-polyethylene glycol thiol (mPEG-SH), as shown schematically in Fig. 6.1. This generates robust covalent bonds to the AuNP surface, providing stability that allows for further application-specific modifications [12]. The modified "place-exchange reaction," as illustrated in Fig. 6.1, provides a paradigm that enables precise tailoring of AuNP properties.

Furthermore, biomolecule functionalization of AuNPs, for example, with antibodies and DNA, and other bio-recognition elements, extends the scope of AuNP utilization for biosensing, imaging, and diagnostic applications. In particular, PEGylation provides further augmentation of biocompatibility, and circulating timesin vivo, and improves pharmacokinetic profiles when applied in the therapeutic setting. Moreover, the use of complex layer-by-layer assembly yields multifunctional coatings, opening a wide range of biomedical and analytical applications [13, 14].

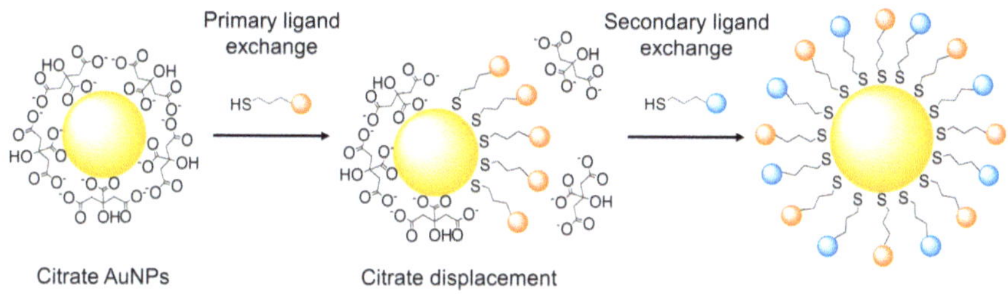

Fig. 6.1 A schematic representation illustrating the modified place-exchange reaction mechanism for ligand modification of AuNPs. Initially, citrate capping agents are replaced with neutral, primary thiol ligands (mPEG-SH), followed by the subsequent introduction of secondary surface-ligands, to achieve well-dispersed AuNPs with precisely controllable surface charges, enhancing their colloidal stability and functional adaptability. Adapted from [15], Fig. 6.1, https://doi.org/10.1038/s41598-019-39.579-3. Licensed under CC BY 4.0 (https://creativecommons.org/licenses/by/4.0/)

6.3 Biocompatibility and Toxicity

AuNPs have greatly advanced biomedical research primary due to their exceptional biocompatibility. Their gold core is inert, which minimizes toxicities to negligible levels; the surface-bound ligands enhances particle stability, and faciltate cellular internalization. However, AuNPs toxicity is influences by factors such as size, shape, surface charge and functionalization, all of which are key parameters. Nevertheless, uncoated and smaller AuNPs have been reported to be cytotoxic effects through the generation of oxidative stress upon interaction with cellular membranes [16]. Conversely, coated AuNPS, particularly those functionalized with PEG, are less toxic, and demonstrate enhanced biocompatibilty. These physicochemical variables must be systematically investigated to enable the optimization of AuNPs for drug delivery, molecular imaging, and therapeutic applications.

6.3.1 In Vitro and in Vivo Studies

An in vitro study investigating the dose-dependent effects of AuNPs on cellular health was recently conducted [17]. At concentrations below 10 μg/mL, AuNPs were generally nontoxic to cells; however, at elevated doses, the cells exhibited damage primarily due to the induction of reactive oxygen species (ROS). Notably, these results underscore the importance of surface modifications, such as PEGylation functionalization, in improving the biocompatibility by minimizing protein corona formation, thereby reducing the likelihood of nonspecific cellular interaction. Particle size also plays a key role, as smaller AuNPs with diameters of approximately 20 nm tend to be retained in tissues. thereby posing the potential for chronic toxicity.

Moreover, the application of certain ligands such as antibodies or peptides to functionalize the AuNP surface provides an enhanced level of targeting, which is particularly critical in therapeutic applications to avoid off-target effects. These studies, when considered collectively, demonstrate a clear requirement for systematic regulation of nanoparticle size, surface properties, and dosage in order to optimize therapeutic effecacy while simultaneously minimizing toxicity.

6.3.2 Influence of Size and Surface Coating on Toxicity of AuNPs

Cytotoxicity associated with the surface interaction of AuNPs with biological systems is significantly influenced by the size and surface modification of AuNPs. When a decrease in the size of AuNPs to dimensions smaller than is observed, evidence suggests a substantial increase in cytotoxicity due to the increased surface area, which promotes the generation reactive oxygen species (ROS), leading to cell membrane disruption, oxidative stress and enhanced penetration of deeper tissues [18].

In contrast, surface modification plays a pivotal role in mitigating these adverse effects. PEGylation is widely recognized for its ability to enhance nanoparticle stability and inhibit nonspecific binding to biological membranes, thereby reducing the formation of protein coronas, which are often implicated immune responses and increased toxicity [19]. Functional groups, such as amines and thiols are employed to improve cellular targeting and reduce nonspecific interactions and their their subsequent cytotoxic consequences.

Additionally, surface coatings serve not only to stabilize the nanoparticles but also to confer additional functional properties. For instance, positively charged AuNPs exhibit stronger electrostatic interactions with negatively charged cell membranes, thereby enhancing cellular uptake and cytotoxicity [29]. Conversely, neutral or negatively charged coatings generally reduce toxicity while maintaining nanoparticle stability. The optimization of size, surface charge, and surface coatings plays an important role in therapeutic applications. Importantly, albumin, or peptide-coated AuNPs have exhibited lower toxicity and improved biocompatibility, highlighting the importance of precise surface engineering for specific biomedical applications.

6.4 Techniques for Functionalization

The surface functionalization of AuNPs is vital for improving colloidal stability, biological compatibility, and multifunctional performance. Ligand exchange reactions are commonly employed to replace weakly bound native capping agents, such as citrate, with thiol-terminated molecules, resulting in the formation of stronger covalent Au–S bonds. The application of precise covalent attachment strategies enables the stable conjugation of biomolecules, such as DNA and proteins, onto the nanoparticle surface. PEG conjugation, often referred to as PEGylation, is widely used to enhance biocompatibility by extending the systemic circulation time in vivo, and by minimizing nonspecific interactions with biological components. Tactics including bioorthogonal strategies such as click chemistry facilitate highly-efficient and selective surface modifications, thereby broadening the functional utility of AuNPs in applications such as targeted drug delivery, biosensing, and molecular imaging.

6.4.1 Ligand Exchange, Covalent Attachment, and Bio-Conjugation

A common method for the functionalization of AuNPs is the exchange of ligands, where weakly adsorbed stabilizers, such as, are gradually exchanged with more strong ligands, such as thiols or phosphines [20]. This process allows the establishment of strong covalent linkages with the ligands which enhances nanoparticle stability and permits site-specific functionalization. For example, thiol-terminated polyethylene glycol (PEG) has been used to prevent aggregation, thereby significantly improving biocompatibility.

Covalent attachment of functional moieties to AuNPs is typically achieved through chemical coupling techniques. Advanced methodologies, such as carbodiimide chemistry, allow the conjugation of either carboxyl or amine groups, facilitating the attachment of biomolecules, including antibodies or peptides [21]. This approach significantly enhances selective targeting and therapeutic efficacy.

Bio-conjugation enriches this framework by incorporating biologically active molecules. For instance, PEGylated AuNPs with ligands. such as aptamers or nucleotides, show promising application in biosensing and diagnostics [22]. Contemporary bio-conjugation strategies, especially click chemistry, provide exceptional selectivity and yields, enabling the development of multifunctional AuNPs in theranostic applications. In total, these methods provide the possibility to tailor properties to meet the specific requirements of diverse applications, ranging from targeted drug delivery systems to environmental monitoring.

6.5 Impact on Applications

The functionalization of AuNPs is an important advancement that has significantly expanded their applicability across multiple domains, including biomedicine, environmental testing, and catalysis. Functionalized AuNPs have been a cornertone in biomedicine for targeted drug delivery. Thus, the presentation of ligands, such as, polyethylene glycol (PEG), or antibodies on their surface enables these nanoparticles to selectively interact with malignant tissues, thereby minimizing off-effects and enhancing therapeutic efficacy [28, 26]. For instance, PEGylated AuNPs provide longer circulation time any biological systems, allowing for better accumulation of the drug at target sites.

Surface-functionalized AuNPs are attractive in the field of biosensing because of their exceptional specificity and responsiveness in highly selective interactions with target analytes, such as DNA and proteins, for LSPR-based diagnostics [27]. This feature has spurred the development of important domain-specific rapid, point-of-care diagnostic portable platforms in the fields of medical and environmental applications, where detection relies on analyte-induced aggregation of the nanoparticles, resulting in visible colorimetric changes [26].

Additionally, their rapid stabilization of active catalytic sites in catalysis increases electron transport and reaction efficiency. This property makes AuNPs viable catalysts for green chemistry, promoting sustainable approaches to pollutant degradation, and energy-efficient chemical processes [28]. Due to their capability to enhance colloidal stability and enable precise surface chemistry, functionalized AuNPs have become a major tool in many cutting-edge technologies, offering novel solutions to emerging problems in healthcare, industry, and environmental sciences. Not only does their versatility demonstrate their transformative power, but also they also represent a critical component in advancing technological innovation and enabling scientific breakthroughs.

6.5.1 Role of Surface Properties in Drug Delivery, Biosensing, and Catalysis

Surface properties are therefore becoming decisive in the administration of drugs, biosensing, and catalysis, as they enable the advancement of innovative platforms to meet important needs in the fields of medicine and environmental science. AuNPs are a clear example of how post-functionalization can greatly enhance their value in these fields.

In drug delivery, functionalized AuNPs exert a significant influence on therapeutic approaches. Biocompatible polymers, such as polyethylene glycol (PEG) , or other targeted ligands, for example, antibodies or aptamers, may be conjugated with AuNPs to confer high specificity toward diseased cells. This specific approach greatly reduces off target effects and improves therapeutic efficiency, and results in the preferential accumulation of drugs at those diseased sites. Some examples include PEGylated AuNPs conjugated with folic acid, which target cancer cells, followed by pH-responsive drug release in the acidic tumor microenvironments [15, 23, 28].

In biosensing, the sensitivity and specificity of AuNP-based sensors are greatly improved by surface modification. These nanoparticles can be functionalized with aptamers or antibodies to enable them to to selectively interact with biomarkers, pathogens, or environmental pollutants. This interaction is often monitored through LSPR, where binding events cause conspicuous changes in color, which are useful in diagnostic applications [8, 27]. Such systems offer real-time detection of critical health indicators, facilitating the development of effective point-of-care diagnostic devices.

Catalysis, however, can be improved through optimal modification of AuNP surfaces with electron donating or accepting groups. It is worth highlighting that these modifications facilitate the stabilization of reaction intermediates and improve electron transfer, thus enhancing the rates of reactions in processes such as CO oxidation and hydrogenation [24]. These enhancements also reinforce the role of AuNPs in supporting green chemistry for sustainable industrial processes.

As such, the concept of surface functionalization can be credited for amplifying the efficacy of AuNPs in drug delivery, biosensing, and catalysis. The relation between surface chemistry, specificity, and stability is not only a breakthrough in current methods, but also a prospect for significant changes in science and technology.

References

1. Feng, J., Gao, J., Yang, W., Liu, R., Shafi, M., Zha, Z.,... Jiang, S.: LSPR optical fiber sensor based on 3D gold nanoparticles with monolayer graphene as a spacer. Opt. Express **30**(6), 10187–10198 (2022).https://doi.org/10.1364/OE.453806
2. Philip, A., Kumar, A.R.: The performance enhancement of surface plasmon resonance optical sensors using nanomaterials: a review. Coordination Chem. Rev. (2022).https://doi.org/10.1016/j.ccr.2022.214424
3. Zhang, P., Hou, H., Xu, S., Wen, Y., Zhang, Y., Xing, F.: Localized surface plasmon resonance sensing based on monometallic gold nanoparticles: from material preparation to detection of bioanalytes. Analyt. Methods (2025). https://doi.org/10.1039/D4AY01509F
4. Karnwal, A., Kumar Sachan, R.S., Devgon, I., Devgon, J., Pant, G., Panchpuri, M.,... Kumar, G.: Gold nanoparticles in nanobiotechnology: from synthesis to biosensing applications. ACS Omega **9**(28), 29966–29982 (2024). https://doi.org/10.1021/acsomega.3c10352
5. Si, P., Razmi, N., Nur, O., Solanki, S., Pandey, C.M., Gupta, R.K.,... de la Zerda, A.: Gold nanomaterials for optical biosensing and bioimaging. Nanoscale Adv. **3**(10), 2679–2698 (2021).https://doi.org/10.1039/d0na00961j
6. Swami, S., Sharma, N., Saini, A.: Captivating nano sensors for mercury detection: a promising approach for monitoring of toxic mercury in environmental samples. RSC Adv. (2024). https://doi.org/10.1039/d4ra02787f
7. Rai, A., Seena, S., Gagliardi, T., Palma, P.J.: Advances in the design of amino acid and peptide synthesized gold nanoparticles for their applications. Adv. Coll. Interface. Sci. **318**, 102951 (2023). https://doi.org/10.1016/j.cis.2023.102951
8. Rosero, W.A.A., Barbezan, A.B., de Souza, C.D., Rostelato, M.E.C.M.: Review of advances in coating and functionalization of gold nanoparticles: from theory to biomedical application. Pharmaceutics **16**(2), 255 (2024). https://doi.org/10.3390/pharmaceutics16020255
9. Lim, S.H., Wong, T.W., Tay, W.X.: Overcoming colloidal nanoparticle aggregation in biological milieu for cancer therapeutic delivery: perspectives of materials and particle design. Adv. Colloid Interface Sci. (2024).https://doi.org/10.1016/j.cis.2024.103094
10. Hock, N., Racaniello, G.F., Aspinall, S., Denora, N., Khutoryanskiy, V.V., Bernkop-Schnürch, A.: Thiolated nanoparticles for biomedical applications: mimicking the workhorses of our body. Adv. Sci. **9**(1), 2102451 (2022). https://doi.org/10.1002/advs.202102451
11. Siegel, A.L., Baker, G.A.: Bespoke nanostars: synthetic strategies, tactics, and uses of tailored branched gold nanoparticles. Nanoscale Adv. (2021).https://doi.org/10.1039/d0na01057
12. Lee, E., Jeon, H., Lee, M., et al.: Molecular origin of AuNPs-induced cytotoxicity and mechanistic study. Sci. Rep. **9**, 2494 (2019). https://doi.org/10.1038/s41598-019-39579-3
13. Choi, J., Kim, B.H.: Ligands of nanoparticles and their influence on the morphologies of nanoparticle-based films. Nanomaterials (2024).https://doi.org/10.3390/nano14201685
14. Hang, Y., Boryczka, J., Wu, N.: Visible-light and near-infrared fluorescence and surface-enhanced Raman scattering point-of-care sensing and bio-imaging: a review. Chem. Soc. Rev. (2022).https://doi.org/10.1039/c9cs00621d
15. Lee, J.W., Choi, S.R., Heo, J.H.: Simultaneous stabilization and functionalization of gold nanoparticles via biomolecule conjugation: Progress and perspectives. ACS Appl. Mater. Interfaces (2021).https://doi.org/10.1021/acsami.1c10436

16. Rezaei, B., Harun, A., Wu, X., Iyer, P. R., Mostufa, S., Ciannella, S.,... Wu, K.: Effect of polymer and cell membrane coatings on theranostic applications of nanoparticles: a review. Adv. Healthcare Mater. **13**(26), 2401213 (2024).https://doi.org/10.1002/adhm.202401213
17. Kus-Liśkiewicz, M., Fickers, P., Ben Tahar, I.: Biocompatibility and cytotoxicity of gold nanoparticles: recent advances in methodologies and regulations. Int. J. Mol. Sci. **22**(20), 10952 (2021). https://doi.org/10.3390/ijms222010952
18. Maghsoudian, S., Motasadizadeh, H., Farhadnejad, H., Fatahi, Y., Nasab, M.H.F., Mahdieh, A.,... Dinarvand, R.: Targeted pH-and redox-responsive AuS/micelles with low CMC for highly efficient sonodynamic therapy of metastatic breast cancer. Biomaterials Adv. **158**, 213771 (2024). https://doi.org/10.1016/j.bioadv.2024.213771
19. Guo, K., Xiao, N., Liu, Y., Wang, Z., Tóth, J., Gyenis, J.,... Shubhra, Q.T.: Engineering polymer nanoparticles using cell membrane coating technology and their application in cancer treatments: Opportunities and challenges. Nano Mater. Sci. **4**(4), 295–321 (2022).https://doi.org/10.1016/j.nanoms.2021.12.001
20. Wang, Y., Bürgi, T.: Ligand exchange reactions on thiolate-protected gold nanoclusters. Nanoscale Adv. (2021)https://doi.org/10.1039/d1na00178g
21. Geng, L., Cui, J., Zhang, C., Yan, Y., Zhao, J., Liu, C.: Chemistry in phase change energy storage: properties regulation on organic phase change materials by covalent bond modification. Chem. Eng. J., 153359 (2024).https://doi.org/10.1016/j.cej.2024.153359
22. Muneer, M.A., Fatima, S., Hussain, N., Mashifana, T., Sayed, A., Boczkaj, G., Rajoka, M.S.R.: Enzyme conjugation-a promising tool for bio-catalytic and biotransformation applications–a review. Topics in Catalysis, 1–17 (2024).https://doi.org/10.1007/s11244-024-01986-w
23. Zhou, B., Guo, X., Yang, N., Huang, Z., Huang, L., Fang, Z.,... Yu, C.: Surface engineering strategies of gold nanomaterials and their applications in biomedicine and detection. J. Mater. Chem. B **9**(28), 5583–5598 (2021).https://doi.org/10.1039/D1TB00181G
24. Zhang, J., Huang, Z., Xie, Y., Jiang, X.: Modulating the catalytic activity of gold nanoparticles using amine-terminated ligands. Chem. Sci. (2022).https://doi.org/10.1039/d1sc05933e
25. Daniel L., Fedlheim Colby A..: Foss (Eds.). Metal nanoparticles: Synthesis, characterization, and applications. CRC Press (2001). https://doi.org/10.1201/9780367800475
26. Aldewachi H., T., Chalati M. N., Woodroofe N., Bricklebank B., Sharrack P.: Gardiner Gold nanoparticle-based colorimetric biosensors Nanoscale **10**(1) 18-33 (2018). https://doi.org/10.1039/C7NR06367A
27. Myalowenkosi, S., Sebastien, B., Mikhael, B., Jean-Marc, J., Krishna, B.: A Review of Gold and Silver Nanoparticle-Based Colorimetric Sensing Assays Advanced Engineering Materials **19**(12) 1700270 (2017).https://doi.org/10.1002/adem.v19.12
28. Rai M., Posten C.: (Eds.). Green biosynthesis of nanoparticles: Mechanisms and applications. CAB International. (2014). https://doi.org/10.1079/9781780642239.0000
29. Oberländer J., Champanhac C., da Costa Marques R., Landfester K., Mailänder V.: Temperature concentration and surface modification influence the cellular uptake and the protein corona of polystyrene nanoparticles Acta Biomaterialia 148271–278 (2022). https://doi.org/10.1016/j.actbio.2022.06.028

Chapter 7
Applications of Gold Nanoparticles

Abstract Gold nanoparticles (AuNPs) are of great interest due to their novel optical, chemical, and biological properties, and have found broad utility in a variety of applications. In biomedicine, AuNPs enhance diagnostic methods, deliver drugs to specific targets, and act as photothermal agents for improving the precision of cancer treatment. The strong surface plasmon resonance (SPR) of AuNPs enables rapid biosensing and real-time detection of diseases. AuNPs are also employed in the environmental applications as catalyst for pollutants degradation, enhancers of air and water purification, and sensors for the sensitive detection of heavy metals and toxic substances. AuNPs are used in nano-fertilizers formulations to stimulate plant growth, in disease diagnostics, and across a range of agriculture applications. Catalysis, flexible electronics, and surface enhanced Raman spectroscopy (SERS) are some of the industrial and technological applications of AuNPs. Based on emerging trends, AuNPs are being integrated into smart materials, nanotheranostics, and hybrid nanostructures, which are representative of future-generation innovations for sustainability, precision medicine, and environmental protection. This chapter underscores the substantial potential of AuNPs across multidisciplinary applications, highlighting their transformative impact on diverse sectors including health, industry, agriculture, and the environment.

7.1 Biomedical Applications

AuNPs are of great interest in biomedical science for their extraordinary optical, electronic, and chemical properties. In addition, the nanoscale size and specific surface properties of AuNPs make them highly biocompatible and enable their application across a wide range of domains, including imaging, diagnostics, and targeted therapies [1]. The study of this fascinating phenomenon of SPR can be used to improve imaging techniques, such as optical coherence tomography, with high sensitivity and spatial resolution for imaging biological species [2]. The high surface area and ease of surface modification have also made AuNPs an essential part of targeted drug delivery systems. They significantly reduce systemic side effects by ensuring

A. C. Alguno et al., *Gold Nanoparticles*, Engineering Materials,
https://doi.org/10.1007/978-981-96-6565-5_7

that the therapeutic action is targeted. For instance, AuNPs can be functionalized with specific antibodies that specifically bind to cancer cells and, thereby direct the treatment t exclusively to diseased cells while preserving healthy tissue.

The versatility of AuNPs through surface modification is evident in their biosensing applications, as they exhibit high sensitivity and specificity in the detection of analytes. Using colorimetric sensing, targeted AuNPs demonstrate changes in SPR related to the biomarker or pathogen presence, enabling real-time detection [1, 2]. Their inherent stability against biologics guarantees reliable and reproducible performance. AuNPs are, in addition, transforming photothermal therapy; if given near-infrared light they can develop localized warmth enough to selectively ablate cancer cells. This unique application exploits the optical properties of AuNPs and opens a new window into non-invasive therapy in cancer treatment.

7.1.1 Anticancer Therapy: Drug Delivery, Photothermal Therapy, and Targeted Therapy

AuNPs are a multifunctional, nanostructured drug delivery system as well as photothermal ablation agent based on the principles of cancer treatment, which exploit the size, physical, optical and thermal properties of the nanoparticles to improve drug bioavailability, photothermal therapy (PTT) efficacy, and site-specific delivery. The efficacy of therapeutic agents, targeting ligands, and biocompatible polymers like PEG attached to the surface of AuNPs make them an efficient drug carrier [3]. The functionalization of AuNPs enhances their circulation time in the bloodstream and facilitates selective accumulation in tumor tissues via the enhanced permeability and retention (EPR) effect, thereby improving the efficacy of treatments, and minimizing invasiveness.

AuNPs exploit SPR of gold to absorb near-infrared (NIR) radiation and induce localized heating, of the particles, that are cytotoxic to cancer cells, with minimal impact on surrounding healthy tissues [4]. Such selective thermal therapies greatly reduce the systemic side effects of conventional cancer treatments and make such therapies more effective, especially in treatment-resistant cancers such as pancreatic cancer (PC) [35]. Shown in Fig. 7.1 is a schematic illustrating AuNPs-induced photothermal treatment in vivo. In summary, AuNPs have proved to be a versatile agent for PTT, as shown in Fig. 7.2. By being able to induce controlled hyperthermia at designated sites, AuNPs possess a significant potential for maximizing treatment efficacy.

The potential of AuNPs is further underscored through their use in targeted cancer therapy. The functionalization of AuNPs with monoclonal antibodies or nucleic acid-based ligands, such as aptamers, has demonstrated remarkable specificity toward tumor-specific biomarkers, including HER2 in breast cancer and EpCAM in gastrointestinal malignancies [6]. This targeted approach minimizes off-target toxicity and has been validated in studies where HER2-targeting AuNP conjugates

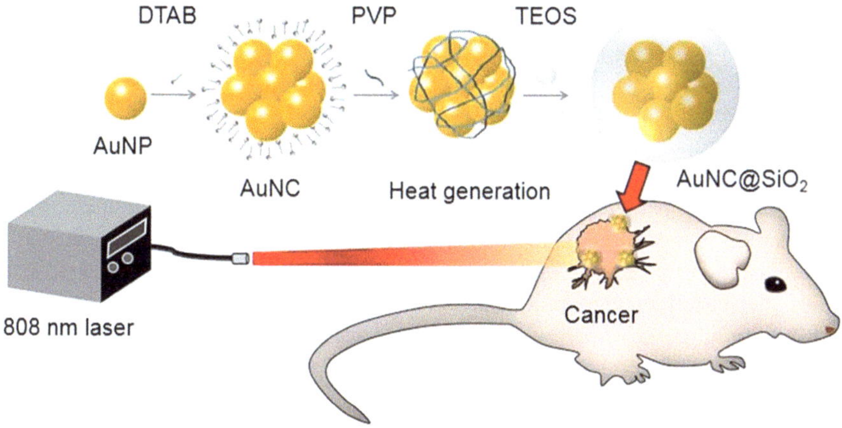

Fig. 7.1 Conceptual diagram of gold nanoparticles-induced photothermal treatment in vivo. To confirm its efficacy, AuNC@SiO_2 was applied to prostate cancers under near-infrared laser irradiation. Adapted from [5], Fig. 7.1, https://doi.org/10.1186/s40580-019-0216-z. Licensed under CC BY 4.0 (https://creativecommons.org/licenses/by/4.0/)

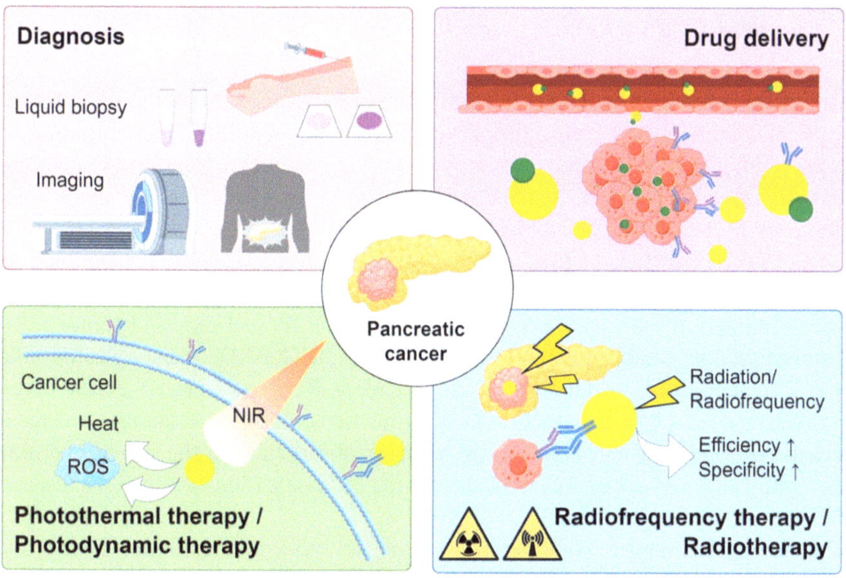

Fig. 7.2 Schematic illustration of gold nanoparticles (AuNPs) in pancreatic cancer (PC) diagnosis and treatment, highlighting their roles in photothermal therapy, drug delivery, and enhanced radiotherapy. AuNPs improve specificity and efficiency in targeting PC, leveraging their NIR and ROS properties while offering eco-friendly, biocompatible, and cost-effective solutions. Adapted from Yin et al. (2024), Fig. 7.2, https://doi.org/10.3390/pharmaceutics16060806. Licensed under CC BY 4.0 (https://creativecommons.org/licenses/by/4.0/)

have been shown to be effective in preclinical breast cancer models [7], indicating a promising direction for precision-based oncologic therapy.

In addition, Fig. 7.2 illustrates the incorporation of AuNPs into imaging-guided therapy systems, facilitating real-time imaging while enhancing the delivery efficiency of therapeutic agents. When combined with theranostic platforms, AuNPs represent a significant advancement, enabling simultaneous diagnostics and therapeutic interventions, thus promoting the realization personalized medicine.

This reflects the intrinsic biocompatibilty and environmentally sustainable nature of AuNPs, thereby reinforcing their growing relevance in clinical oncology. These trends showcase their tremendous promise in combating the challenges associated with highly aggressive malignancies, such as pancreatic cancer (PC), particularly through SPR-mediated phototherapy therapy and radio-sensitized nanocarrier drug delivery systems. AuNPs remain at the forefront of these innovations as they enable targeted, efficient, and personalized strategies in cancer treatment and treatment response.

7.1.2 Diagnostics: Biosensors, Imaging Agents, and Biomarkers

AuNPs are advancing in the application of medical diagnostics due to their distinct physical, chemical, and optical properties. The application of SPR has significantly improved the sensitivity and specificity of various diagnostic platforms, including sensors, imaging systems, and biomarker assay, thus enhancing the quality and efficiency of healthcare delivery [8–10].

AuNP-based biosensors have revolutionized the detection of biological molecules, such as DNA, proteins, and glucose. Because they intrinsically amplify signals innately, these biosensors serve as critical tools for real-time and point-of-care diagnostics. One notable application has been in rapid COVID-19 testing using SPR-enhanced techniques to identify RNA-dependent RNA polymerase genes. These biosensors yield a visible message via SPR-induced colorimetric changes as a result of analyte aggregation, which can be visually observed with the naked eye, thus presenting an easily accessible indicator for the early detection of infectious diseases [2].

In **medical imaging**, AuNPs find their major application in improving various techniques, including computed tomography (CT) and molecular imaging [11]. They can be conjugated to targeting ligands on their surface to allow for targeted accumulation at specific sites, which improves the accuracy of imaging for cancers and cardiovascular diseases. Moreover, this functionalization not only enhances contrast, but also deepens understanding of cellular events, and therefore improves the diagnosis and treatment of diseases at the early stages [12].

For biomarkers, AuNPs provide exceptionally high sensitivity through by the SPR-based colorimetric assay. Coupling AuNPs to antibodies or aptamers can detect

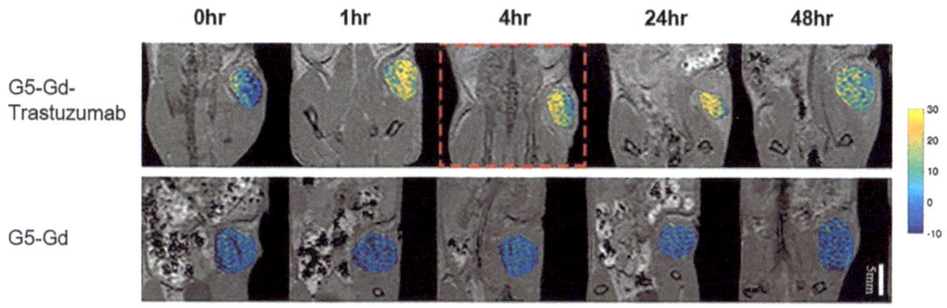

Fig. 7.3 MRI images of tumors in mice injected with either G5-Gd-Trastuzumab or G5-Gd nanoparticles. Imaging was performed before injection (baseline) and at 1, 4, 24, and 48 hours post-injection. Adapted from [15], Fig. 7.3, https://doi.org/10.1186/s12951-020-00695-2. Licensed under CC BY 4.0 (https://creativecommons.org/licenses/by/4.0/)

biomarkers at extremely low concentrations, such as the prostate-specific antigen (PSA) [13]. They are particularly suitable for studies on disease pathogenesis, as protein biomarkers contribute greatly to refining diagnostic specificity, and thus biomarkers can provide necessary information for effective disease profiling and management. The trend is moving toward multiplex assays, to enable healthcare professionals to analyze a greater number of biomarkers simultaneously [14].

Recently developed wearable health gadgets have incorporated AuNPs. For instance, colorimetric strips are being developed to monitor glucose and sweat biomarkers non-invasively. In environmental health, SPR based technology is used to identify pollutants, thereby bridging diagnostics and public health [9].

These breakthroughs are transformative in diagnostics, leading to earlier diagnosis, personalized medicine, and wider availability of health vigilance. Real-time monitoring systems are being improved for accuracy through the combination of nanotechnology with artificial intelligence (AI) for data interpretation. Research is being conducted to make these technologies more affordable and sustainable, thereby increasing access for underserved populations [1]. As shown in Fig. 7.3, the AuNP conjugate is capable of efficiently targeting and imaging human epidermal growth factor receptor 2 (HER-2) positive tumors in vivo, which is a starting point for the use of this diagnostic tool for early detection, metastatic assessment, and therapeutic monitoring of HER-2 positive cancers.

7.2 Environmental Applications

Applications of AuNPs in the detection and purification of pollutants and catalysis are revolutionizing the field of environmental science. Due to their high surface area and unique SPR properties, these nanomaterials have garnered significant interest for detecting hazardous heavy metal and organic pollutants. AuNP-based colorimetric sensors provide rapid, highly sensitive, visual detection of toxic agents in water, and their catalytic properties are capable of degrading environmental toxins, thus contributing to healthier ecosystems. Moreover, functionalized AuNPs can detect and monitor the health of the environment in real-time, which facilitates the sustainable management of ecosystems, ultimately leading to enhanced environmental cleanliness.

7.2.1 Air and Water Purification

AuNPs have been found to be excellent catalysts and adsorbents in the treatment of air and water pollution. In water treatment, their functionalized forms are highly effective in removing a broad spectrum of pollutants from water including metals, dyes and organic matter. Among these, AuNPs coated with thiol or amine groups have a very high tendency to react with heavy metals, such as mercury and lead, and hence can be used to remove them from contaminated bodies.

In the area of air quality, AuNPs serve as catalysts in oxidation processes that break up toxic gases, including carbon monoxide (CO) and volatile organic compounds (VOCs) [16]. They enhance the rate of reactions at lower temperatures, which is highly energy-efficient, and therefore very beneficial. AuNPs further find application in advanced air purifiers and sensing devices for trapping and neutralizing gaseous contaminants. They are used in purification systems for long durations due to their capacity for regeneration and extreme stability under adverse conditions. As shown in Fig. 7.4, a representative extraction procedure demonstrates the utility of functionalized AuNPs in efficient water purification.

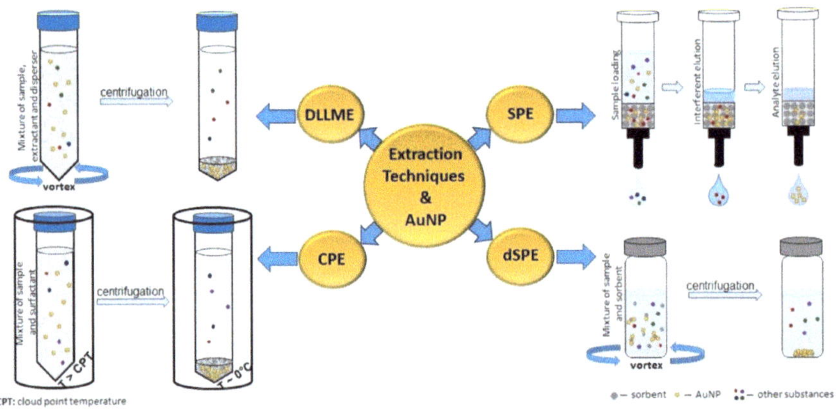

Fig. 7.4 Schematic illustration of four recommended extraction techniques, dispersive liquid–liquid microextraction (DLLME), solid-phase extraction (SPE), dispersive solid-phase extraction (dSPE), and (CPE), for the selective separation and preconcentration of gold nanoparticles (AuNPs) from complex environmental matrices. Adapted from [18], Fig. 7.4, https://doi.org/10.3390/ijms231911465. Licensed under CC BY 4.0 (https://creativecommons.org/licenses/by/4.0/)

Recent studies, however, affirmed a synergistic effect of AuNPs with other nanostructured materials, such as e graphene, for achieving better adsorption and catalytic activity of the system [17]. As a result, this hybrid strategy not only improves the extraction of various pollutant but also reduces the operation cost, making this environmental cleaning strategy both eco-friendly and sustainable.

7.2.2 Detection of Environmental Pollutants

AuNPs are a significant improvement in environmental pollution detection and remediation due to their size dependent SPR-based colorimetric properties [13]. Functionalized AuNPs are highly sensitive to changes in the environment and are therefore particularly effective to detect traces of heavy metals, pesticides, and organic compounds in the environment. For example, thiol modified AuNPs have a distinct colorimetric change on the addition of mercury ions, wherein the NPs change from red to blue [36].

These novel colorimetric sensors are not only economical, but also suitable for in-situ measurement, which is very advantageous for real-world environmental conditions. The ability to perform simultaneous analysis of several analytes, for instance, arsenic and cadmium ions, in a single assay is an enormous advantage in analyzing environmental samples.

Beyond detection, AuNPs are also effective in remediating pollutants. They also have a catalytic role in the degradation of persistent organic pollutants (POPs) such as polychlorinated biphenyls (PCBs). Furthermore, AuNPs improve the rate of photocatalytic processes, which lead to the decomposition of toxic substances [37]. A good example of this is shown in Fig. 7.4, utilizing AuNPs for mercury detection. This multifunctional role therefore makes AuNPs crucial in the improvement of environmental analysis and treatment, thereby contributing to the achievement of global sustainability goals. Their pioneering applications show the potential of nanoparticle technology for the protection of the environment in the future (Fig. 7.5).

7.3 Agricultural Applications

AuNPs are transforming the practice of agriculture by enhancing precision and sustainability. The current review highlights their numerous applications in nanosensors for real-time assessment of soil and plant conditions, targeted delivery of agrochemicals, and detection of pathogens. AuNPs decrease the environmental impact by improving control over the release of fertilizers and pesticides. They also incorporate optical and electrochemical sensing systems for the evaluation of plant nutrient and stress status. Active components of AuNP-based nanosensors take advantage of LSPR to detect heavy metals and soil contaminants, thus contributing to sustainable farming and increased yield with lower ecological impact.

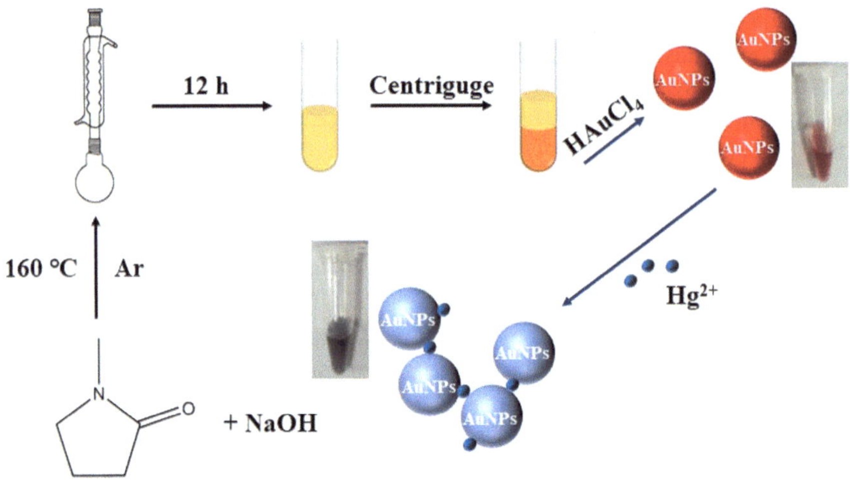

Fig. 7.5 Schematic illustration of the synthesis of gold nanoparticles (AuNPs) using oxidized N-methylpyrrolidone (NMP*) under alkaline and high-temperature conditions, followed by their application in a colorimetric assay for Hg^{2+} detection. Adapted from [19], Fig. 7.5, https://doi.org/10.1038/s41598-023-49551-x. Licensed under CC BY 4.0 (https://creativecommons.org/licenses/by/4.0/)

7.3.1 Role in Plant Growth and Seed Germination

Reference [20] demonstrated that AuNPs modulate physiological and biochemical processes that are essential for plant growth and seed germination. They facilitate the absorption of water and nutrients by interacting directly with plant cells and enhancing photosynthesis activity. AuNPs possess the capacity to penetrate seed coatings and cell membranes, thereby enabling the delivery of essential macronutrients such as nitrogen (N) and phosphorus (P) to metabolically active zones, which subsequently stimulate key enzymatic activities necessary for successful germination [21]. The catalytic functionalities of these nanoparticles serve to reduce reactive oxygen species (ROS), thereby mitigating oxidative oxidative stress and reinforcing seedling vigor.

According to Palacios-Márquez et al. (2023) [34], surface-functionalized AuNPs exhibit promising utility as nano-biofortification systems capable of delivering essential micronutrients to crops. This function includes enhancement of the plant's antioxidant defense mechanisms, thereby improving tolerance to abiotic stressors such as drought and salinity. Empirical data from field trials have indicated that AuNP treatment promotes accelerated and synchronized germination, augmented root and shoot elongation, and elevated biomass production [22, 23] underscoring their potential as transformative agents in the agricultural sector.

The inherent biocompatibility of AuNPs and their behavior as nanozymes across multiple biochemical pathways present significant opportunities for advancing agricultural biotechnology. However, precaution must be taken to avoid cytotoxic effects on plant tissues, necessitating the optimization of nanoparticle concentration and exposure parameters. Therefore, the strategic incorporation of AuNPs in plant science holds considerable promise for redefining future directions in sustainable agricultural and global food security.

7.3.2 *Impact on Crop Yield and Stress Tolerance*

AuNPs are a significant innovation in agricultural science that enhance crop yield and stress tolerance modulating gene expression and cellular metabolism at the molecular level. AuNPs engineered are nanosystems designed to enhance the delivery of growth regulators, nutrients, and pesticides, thereby increasing their bioavailability and assimilation [24]. By elevating photosynthetic efficiency and nitrogen assimilation, this nanointervention leads to enhanced biomass accumulation and overall crop productivity. Through the upregulation of antioxidant enzymes such as superoxide dismutase and catalase [25], AuNPs protect plants from abiotic and biotic stressors, including drought, salinity, and pathogenic infections. They further regulate hormonal networks, including those of abscisic acid and gibberellins, which are key mediators in stress perception and adaptative responses.

In addition, AuNPs enhance water retention and facilitate nutrient transport within plant tissues, thereby mitigating the negative consequences of environmental stressors [26, 27]. They can also be integrated in precision agriculture systems to prevent disease outbreaks and pests infestations through advanced nanosensors and targeted delivery platforms. However, while the agronomic advantages of AuNPs are noteworthy, it is imperative that their environmental impact and biosafety profiles be investigated with scientific rigor. These challenges must be addressed comprehensively before AuNP technology can be safely and responsibly incorporated into conventional agricultural systems, ultimately paving the way for sustainable, resilient, and climate-adaptive production strategies.

7.4 Industrial and Technological Applications

AuNPs are used in a wide array of scientific and industrial fields because of their unique optical, electronic and catalytic properties. In electronics, they contribute to the advancement of flexible printed circuits by serving as components in conductive inks, thereby facilitating device miniaturization and enhancing overall efficiency. They are important in accelerating reactions in the energy and environmental sectors, where their catalytic capabilities are considered indispensable. AuNPs play an important role in SERS enabling the ultrasensitive detection of trace amounts of analytes for applications in chemical sensing, environmental monitoring and clinical diagnostics. Furthermore, their application extends beyond the biomedical domain, encompassing both targeted drug delivery systems and advanced diagnostic platforms, and therefore holds significant promise for broad industrial integration.

7.4.1 Catalysis, Electronics, and Surface-Enhanced Raman Spectroscopy (SERS)

AuNPs are currently being applied in both heterogeneous catalysis and diverse nanotechnological applications [28] and exhibit exceptional performance in oxidation and reduction reactions due to their high surface area and abundance of catalytically active sites. They are most active in carbon monoxide oxidation and chemoselective hydrogenation reactions, and thus serve as indispensable catalytic agents in green chemistry. Furthermore, they possess superior electrical conductivity and excellent compatibilty with nanoscale device fabrication, rendering them valuable for the development of new sensors, transistors and memory devices.

One of the most powerful applications of AuNPs is surface-enhanced Raman spectroscopy (SERS), which is based on the LSPR concept. This enhances Raman signals to detect trace amounts of analytes, which is important in critical areas such as environmental monitoring, food safety, and biomedical diagnostics. The ability of AuNPs to amplify the intensity of molecular vibration when biomolecules or pollutants are absorbed onto the surface is an important aspect in the application of SERS [29]. In addition, strategic functionalization with specific ligands allows for the selective and sensitive detection of target analytes. The multifunctional capabilities of AuNPs clearly illustrate their transformative potential, marking the beginning of a new era in technology and scientific research, and positioning them to contribute significantly to a wide range of fields.

7.5 Future Directions for AuNP Applications

A transformative era of AuNP applications is anticipated to be based on the conjunction of smart technologies and smart practices is rapidly emerging. Advanced methods for green synthesis and innovative hybrid nanostructures are being developed, exhibiting enhanced physicochemical properties through the coupling of AuNPs with other functional nanomaterials. AuNPs are poised to revolutionize precision agriculture through real-time monitoring strategies and targeted delivery systems. They enable the implementation of personalized therapies and non-invasive diagnostics in nanomedicine. Further, their role in renewable energy, especially in enhancing photovoltaic efficiency, holds significant promise. However, to fully harness their potential, challenges related to scalability, environmental impact, and biocompatibility must be systematically addressed.

7.5.1 Smart Materials, Responsive Nanoparticles, and Hybrid Structures

AuNPs in the development of smart materials represents the current trend in many sectors due to their incredible versatility and responsive nature. These stimuli-receptive ligands within the nanostructures enable unprecedented control over the active or passive properties of the nanoparticles in response to various environmental stimuli such as pH, temperature, light [30]. This adaptability underpins innovations in personalized drug delivery and responsive biosensors, which are expected to enhance medical diagnosis precision.

Hybrid architectures that combine AuNPs with graphene and polymers significantly enhance their mechanical, optical, and electrical properties [31]. These hybrid composites are enabling groundbreaking applications, such as flexible electronics, energy storage devices and catalytic processes. The emergence of novel AuNP-based metal hybrids demonstrates considerable potential in supercapacitor and fuel cell applications, for example, the excellent alternative provided through the interface of AuNPs and graphene.

Hybrid nanostructures facilitate more precise and effective pollutant detection and removal methods in environmental science, significantly benefiting ecosystem health. Additionally, functionalized AuNPs are at the forefront of nanotheranostics, which integrate diagnostic and therapeutic functions on the same platform, aiming to address complex diseases like cancer [32, 33]. The advancement of programmable AuNP systems is enabling the development of self-healing and adaptive materials, which is a major step forward in the direction of sustainability. These innovations align with worldwide sustainable improvement, both by improving energy efficiency and optimizing resource utilization, thus advancing responsible and intelligent materials science.

References

1. Rai, M., Posten, C. (Eds.): Green biosynthesis of nanoparticles: Mechanisms and applications. CABI (2013). https://doi.org/10.1079/9781780642239.0000
2. Sabela, M., Balme, S., Bechelany, M., Janot, J.-M., Bisetty, K.: A review of gold and silver nanoparticle-based colorimetric sensing assays. Adv. Eng. Mater. (2017).https://doi.org/10.1002/adem.201700270
3. Hossain, A., Rayhan, M.T., Mobarak, M.H., Rimon, M.I.H., Hossain, N., Islam, S., Al Kafi, S.A.: Advances and significances of gold nanoparticles in cancer treatment: a comprehensive review. Results Chem., 101559 (2024).https://doi.org/10.1016/j.rechem.2024.101559

4. Khurana, K., Jaggi, N.: Localized surface plasmonic properties of Au and Ag nanoparticles for sensors: a review. Plasmonics (2021).https://doi.org/10.1007/s11468-021-01381-1
5. Kim, J., Chun, S.H., Amornkitbamrung, L., Song, C., Yuk, J.S., Ahn, S.Y., Kim, B.W., Lim, Y.T., Oh, B.-K., Um, S.H.: Gold nanoparticle clusters for the investigation of therapeutic efficiency against prostate cancer under near-infrared irradiation. Nano Convergence **7**(5) (2020). https://doi.org/10.1186/s40580-019-0216-z
6. Deng, G., Zha, H., Luo, H., Zhou, Y.: Aptamer-conjugated gold nanoparticles and their diagnostic and therapeutic roles in cancer. Frontiers Bioeng. Biotechnol. **11**, 1118546 (2023). https://doi.org/10.3389/fbioe.2023.1118546
7. Omidi, Y., Mobasher, M., Castejon, A.M., Mahmoudi, M.: Recent advances in nanoscale targeted therapy of HER2-positive breast cancer. J. Drug Target. **30**(7), 687–708 (2022). https://doi.org/10.1080/1061186X.2022.2055045
8. Feldheim, D.L., Foss, C.A., Jr. (Eds.).: Metal nanoparticles: Synthesis, characterization, and applications. Marcel Dekker (2002)
9. Sadiq, Z., Safiabadi Tali, S.H., Hajimiri, H., Al-Kassawneh, M., Jahanshahi-Anbuhi, S.: Gold nanoparticles-based colorimetric assays for environmental monitoring and food safety evaluation. Crit. Rev. Anal. Chem. **54**(7), 2209–2244 (2024). https://doi.org/10.1080/10408347.2022.2162331
10. Tai, J., Fan, S., Ding, S., Ren, L.: Gold nanoparticles based optical biosensors for cancer biomarker proteins: a review of the current practices. Frontiers Bioengineering Biotechnol. **10**, 877193 (2022). https://doi.org/10.3389/fbioe.2022.877193
11. Wang, A., Qi, W., Gao, T., Tang, X.: Molecular contrast optical coherence tomography and its applications in medicine. Int. J. Molecular Sci. (2022). https://doi.org/10.3390/ijms23063038
12. Niederberger, M., Pinna, N.: Metal Oxide Nanoparticles in Organic Solvents: Synthesis, Formation, Assembly and Application. Springer (2009).https://doi.org/10.1007/978-1-84882-671-7
13. Karnwal, A., Kumar Sachan, R.S., Devgon, I., Devgon, J., Pant, G., Panchpuri, M.,... Kumar, G.: Gold nanoparticles in nanobiotechnology: from synthesis to biosensing applications. ACS Omega **9**(28), 29966–29982 (2024). https://doi.org/10.1021/acsomega.3c10352
14. Luque, R., Varma, R.S. (Eds.): Sustainable preparation of metal nanoparticles: methods and applications. Royal Soc. Chem. (2012)
15. Chen, J.S., Chen, J., Bhattacharjee, S., Cao, Z., Wang, H., Swanson, S.D., Zong, H., Baker, J.R., Wang, S.H.: Functionalized nanoparticles with targeted antibody to enhance imaging of breast cancer in vivo. J. Nanobiotechnol. **18**, 135 (2020). https://doi.org/10.1186/s12951-020-00695-2
16. Kim, H.S., Kim, H.J., Kim, J.H., Kim, J.H., Kang, S.H., Ryu, J.H.,... Bae, J.W.: Noble-metal-based catalytic oxidation technology trends for volatile organic compound (VOC) removal. Catalysts **12**(1), 63 (2022). https://doi.org/10.3390/catal12010063
17. Ventura-Espinosa, D., Martín, S., García, H., Mata, J.A.: Ligand effects in the stabilization of gold nanoparticles anchored on the surface of graphene: implications in catalysis. J. Catalysis (2021). https://doi.org/10.1016/j.jcat.2020.12.027
18. Hagarová, I., Nemček, L., Šebesta, M., Zvěřina, O., Kasak, P., Urík, M.: Preconcentration and separation of gold nanoparticles from environmental waters using extraction techniques followed by spectrometric quantification. Int. J. Mol. Sci. **23**(19), 11465 (2022). https://doi.org/10.3390/ijms231911465
19. Shao, X., Yang, D., Wang, M. et al.: A colorimetric detection of Hg^{2+} based on gold nanoparticles synthesized using oxidized N-methylpyrrolidone as a reducing agent. Scientific Reports **13**, 22208 (2023).https://doi.org/10.1038/s41598-023-49551-x
20. Sakthivel, A., Chandrasekaran, R., Balasubramaniam, S., Sathyanarayanan, H., Gnanajothi, K.: Nanomaterials as potential plant growth modulators: applications, mechanism of uptake, and toxicity: a comprehensive review. BioNanoScience **15**(1), 1–21 (2025). https://doi.org/10.1007/s12668-024-01648-x
21. Zaim, N.S.H.B.H., Tan, H.L., Rahman, S.M.A., Abu Bakar, N.F., Osman, M.S., Thakur, V.K., Radacsi, N.: Recent advances in seed coating treatment using nanoparticles and nanofibers for

enhanced seed germination and protection. J. Plant Growth Regul. **42**(12), 7374–7402 (2023). https://doi.org/10.1007/s00344-023-11038-4
22. Agrawal, S., Kumar, V., Kumar, S., Shahi, S.K.: Plant development and crop protection using phytonanotechnology: a new window for sustainable agriculture. Chemosphere (2022).https://doi.org/10.1016/j.chemosphere.2022.134465
23. Jampílek, J., Kráľová, K.: Nanoparticles for improving and augmenting plant functions. In: Advances in Nano-Fertilizers and Nano-Pesticides in Agriculture, pp. 171–227. Woodhead Publishing (2021). https://doi.org/10.1016/B978-0-12-820092-6.00008-2
24. Hassanisaadi, M., Barani, M., Rahdar, A., Heidary, M., Thysiadou, A., Kyzas, G.Z.: Role of agrochemical-based nanomaterials in plants: Biotic and abiotic stress with germination improvement of seeds. Plant Growth Regul. **97**(2), 375–418 (2022). https://doi.org/10.1007/s10725-021-00782-w
25. Timoszyk, A., Grochowalska, R.: Mechanism and antibacterial activity of gold nanoparticles (AuNPs) functionalized with natural compounds from plants. Pharmaceutics (2022).https://doi.org/10.3390/pharmaceutics14122599
26. Thabet, S.G., Alqudah, A.M.: Unraveling the role of nanoparticles in improving plant resilience under environmental stress condition. Plant and Soil (2024).https://doi.org/10.1007/s11104-024-06581-2
27. Wahab, A., Batool, F., Muhammad, M., Zaman, W., Mikhlef, R.M., Naeem, M.: Current knowledge, research progress, and future prospects of phyto-synthesized nanoparticles interactions with food crops under induced drought stress. Sustainability **15**(20), 14792 (2023). https://doi.org/10.3390/su152014792
28. Hassanisaadi, M., Bonjar, G.H.S., Rahdar, A., Pandey, S., Hosseinipour, A., Abdolshahi, R.: Environmentally safe biosynthesis of gold nanoparticles using plant water extracts. Nanomaterials **11**(8), 2033 (2021). https://doi.org/10.3390/nano11082033
29. Kalashgrani, M.Y., Mousavi, S.M., Akmal, M.H., Gholami, A., Omidifar, N., Chiang, W.H.,... Rahman, M.M.: Gold fluorescence nanoparticles for enhanced SERS detection in biomedical sensor applications: current trends and future directions. Chem. Record, e202300303 (2024). https://doi.org/10.1002/tcr.202300303
30. Lavrador, P., Esteves, M.R., Gaspar, V.M., Mano, J.F.: Stimuli-responsive nanocomposite hydrogels for biomedical applications. Adv. Func. Mater. **31**(8), 2005941 (2021). https://doi.org/10.1002/adfm.202005941
31. Sanchis-Gual, R., Coronado-Puchau, M., Mallah, T., Coronado, E.: Hybrid nanostructures based on gold nanoparticles and functional coordination polymers: chemistry, physics and applications in biomedicine, catalysis and magnetism. Coord. Chem. Rev. **480**, 215025 (2023). https://doi.org/10.1016/j.ccr.2023.215025
32. Al-Thani, A.N., Jan, A.G., Abbas, M., Geetha, M., Sadasivuni, K.K.: Nanoparticles in cancer theragnostic and drug delivery: a comprehensive review. Life Sci., 122899 (2024). https://doi.org/10.1016/j.lfs.2024.122899
33. Kashyap, B.K., Singh, V.V., Solanki, M.K., Kumar, A., Ruokolainen, J., Kesari, K.K.: Smart nanomaterials in cancer theranostics: challenges and opportunities. ACS Omega **8**(16), 14290–14320 (2023). https://doi.org/10.1021/acsomega.2c07840
34. Palacio-Márquez, A., Ramírez-Estrada, C.A., Sánchez, E., Chávez-Mendoza, C.: Biofortification with zinc: the use of nanoparticles as an alternative to improve the nutritional quality of crops. In: Advanced Nanotechnology in Plants (1st ed., p. 17). CRC Press (2023). https://doi.org/10.1201/9781003368571
35. Rongguan, Y., Jirameth, T., Akhtar, G., Jaepil, J., Xiaolei, H., Yuqi, Z., Hanshu, W., Qiqi, L., George, F., Alamgir, K., Michael, R., Bockstaller, K., M.: Organic nanoparticles with tunable size and rigidity by hyperbranching and cross-linking using microemulsion ATRP. Proceedings of the National Academy of Sciences. **121**(29) (2024).https://doi.org/10.1073/pnas.2406337121
36. Chao, W., Chenxu, Y.: Detection of chemical pollutants in water using gold nanoparticles as sensors: a review Abstract Reviews in Analytical Chemistry. **32**(1) 1–14 (2013). https://doi.org/10.1515/revac-2012-0023

37. Juan, Q., Luwen, Q., Juntong, Z., Yiqing, Z., Jian, S., Jinyou, S., Changjin, O.: Accelerated anaerobic biodecolorization of sulfonated azo dyes by magnetite nanoparticles as potential electron. transfer mediators Chemosphere. **263** 128048 (2021). https://doi.org/10.1016/j.chemosphere.2020.128048

Chapter 8
Challenges and Future Perspectives in Aunp Research

Abstract Gold nanoparticles (AuNPs) are currently being utilized across medicine, electronics, and environmental sciences, but there are several critical challenges that prevent their optimal use. Inconsistencies in synthesis lead to variability in the size, shape and stability of the nanoparticles, which in turn compromises the reproducibility and scalability. Environmental concerns are exacerbated by the use of toxic chemicals in conventional synthesis protocols, which calls for a change to green, sustainable sustainable synthesis. Safety, handling, and disposal are some of the unresolved barriers hindering clinical translation, and for that matter there is a need for uniform, internationally accepted standardized guidelines. The main drawback is that laboratory techniques often fail to meet the requirements of industry. Some of the emerging approaches include microfluidic synthesis, artificial intelligence-driven optimization, and hybrid nanomaterials, which offer improved control and production scalability. However, in healthcare, AuNPs are transforming diagnosis, drug delivery, and cancer treatment; nonetheless, cost remains a significant issue. The future generation of smart materials, quantum dots, and flexible electronics will certainly enhance their role. These challenges can be met through interdisciplinary collaboration, and robust ethical consideration, to enable the application of AuNPs in transforming science, industry, and sustainability on a global scale.

8.1 Current Challenges in AuNP Research

Research into gold nanoparticles (AuNPs) is extensive and of considerable significance, but it has many challenges that must be overcome in order to realize its full potential. One of the main issues is the lack of consistency in synthesis method, which leads to variability in size, shape and functionalization—all of which affect the reliability of AuNPs in catalysis and biomedical applications [1]. Another problem is that the output of the technology must be significantly scaled to meet industrial demand, and the precision in size and shape control must surpass current values by at least one order of magnitude. The environmental impact of the synthesis is also a concern due to the use of toxic precursors and the generation of toxic byproducts, in addition

A. C. Alguno et al., *Gold Nanoparticles*, Engineering Materials,
https://doi.org/10.1007/978-981-96-6565-5_8

to the fact that there are no standard operating procedures as a consequence of the nascent stage of this chemical domain [2]. In order to move forward, green synthesis techniques should be supported, quality control measures should be improved, and interdisciplinary collaboration should be encouraged. Thus, resolving these multi-faceted issues will empower the scientific community to facilitate the transition of AuNPs into scalable, real-world applications, and thus ensure their significance for the development of new technologies and enhancement of global human well-being.

8.1.1 Environmental Impact of Chemical Synthesis

The conventional fabrication of AuNPs has usually relied on toxic reducing agents, including sodium borohydride, that produce hazardous by-products which pose serious environmental hazards [3] These traditional chemical processes are also unsustainable for large-scale applications and also pollute the environment. The management of these by-products particularly in aqueous media can be detrimental to the environment and pose a threat to aquatic life, and may lead to the incorporation of toxic substances in the food chain. Moreover, these chemical methodologies are often energy-intensive, thereby exacerbating carbon emissions and contributing to global climate change.

However, green synthesis has emerged as a viable and environmentally benign alternative. This approach uses plant extracts, bacteria, and fungi as eco-friendly reducing and stabilizing agents [4, 5]. This innovative approach typically enhances energy efficiency while substantially minimizing environmental damage. It significantly reduces ecological toxicity and demonstrates superior energy efficiency in comparison to conventional methods. However, there are some issues concerning the control over the size, morphology and functionality of nanoparticles generated by green synthesis methods.

In order to move forward towards a sustainable future, it is crucial to design scalable and economical environmentally friendly approaches [6, 7] that decrease the ecological impacts of AuNP manufacturing by a significant margin. This demands the cooperation of material chemists, environmental researchers, and regulatory authorities, and nanotechnologists to define standard operating procedures and perform life-cycle analysis. Such interdisciplinary partnerships will be essential for the transition to environmentally responsible practices in nanotechnology, and will be critical in fostering award-winning advances in scientific discovery and innovation.

8.1.2 Issues of Reproducibility and Scalability

Reproducibility and scalability are the primary challenges in AuNPs research and commercialization. Small changes in synthesis conditions such as temperature, pH, and reagent concentration result in variations in size, shape and surface properties of the nanoparticles. This variability renders AuNPs unreliable across different applications, ranging from diagnostics to energy solutions. Laboratory-scale synthesis techniques often fail to translate effectively into industrial-scale production, where uniformity is a key issue. Aggregates, batch to batch variation, and the management of reaction kinetics management are additional challenges in the context of large-scale manufacturing.

In order to overcome these challenges, it is crucial to establish standard synthesis conditions and to use more sophisticated manufacturing methods, such as microfluidics, which provide continuous and more accurate production of nanoparticles [8, 9]. An online monitoring step integrated with automation can improve the control of the synthesis [10]. Furthermore, using computational modelling to predict the behavior of the reaction system and kinetics may also help improve reproducibility.

Such collaboration between academic research and industrial practice will be essential to addressing these challenges and enhancing the industrial applicability of AuNP technologies. Hence, this elucidate multi-faceted strategies to support the scientific community in promoting stable and scalable AuNPs, and consequently amplify the potential applicability of AuNP technologies in diverse domains [11].

8.2 Regulatory and Ethical Considerations

The AuNPs' regulatory and ethical landscape is a highly fragmented and evolving regulatory environment between countries. The REACH regulation (Registration, Evaluation, Authorization, and Restriction of Chemicals) is the main driver for nanoparticle regulation in the European Union; under this regulation all nanoparticles must be tested for toxicity and meet rigorous safety standards before they can be released into the environment [12]. In contrast, the United States uses the Toxic Substances Control Act (TSCA), which is administered by the Environmental Protection Agency (EPA), to regulate the reporting of chemicals and the measures to be taken in the management of chemicals including the disposal of nanoparticles [13].

As nanomedicine advances, ethical considerations become more critical and thus, there is the need to strictly follow the principles of informed consent in clinical trials in order to protect the patient and their autonomy [14]. For instance, India has very strict rules and regulations in this domain through the Indian Council of Medical Research (ICMR), which stipulates that all nanomedicines must undergo thorough preclinical toxicity tests before being moved to human clinical trials [15]. In addition, issues regarding the availability of nanoparticle-based technologies in healthcare are also of great concern. The high costs of nanomedicine disproportionately benefit affluent populations and exacerbate socio-economic disparities in health outcomes [16].

Multi-stakeholder collaboration is essential to harmonize global norms and procedures. The Testing and Assessment Programme on Manufactured Nanomaterials (TPMN) of the OECD is an excellent example of this cooperation, which provides a common ground for the development of guidelines for testing manufactured nanomaterials in order to protect human and environmental health [13]. Ultimately, clear risk communication is important, as is ethical manufacturing and disposal. It will be necessary to re-evaluate and adapt existing regulatory frameworks in light of emerging developments to ensure the long-term sustainability of innovation in nanoparticle deployment.

8.2.1 Guidelines for Safe Handling and Disposal

The proper handling and disposal of AuNPs is a major issue that concerns researchers, end users and environmental stakeholders. It is critically necessary to follow the safety guidelines which include the use of personal protective equipment (PPE) such as gloves, laboratory coats and N95 respirators in situations where nanoparticle aerosols may be produced. The other factor that cannot be overemphasized is the need to have proper laboratory ventilation systems such as fume hoods to prevent exposure to airborne nanoparticles. Moreover, AuNPs should be kept in properly labelled, closed containers, in order to avoid interactions with other substances, which may cause a hazardous reaction [12, 14].

At the moment, there is no single global standard for the disposal of AuNPs across jurisdictions. For example, the REACH regulation of the European Union is responsible for providing detailed information on the toxicity and environmental impact of nanomaterials, and for encouraging the appropriate waste management options such as incineration and chemical fixation [17, 13]. In the United States, the EPA has set forth guidelines under the Resource Conservation and Recovery Act (RCRA) that classify nanoparticle waste as hazardous, which must be disposed of in designated facilities [18]. In contrast, India has very strict guidelines for the pre-treatment of AuNP waste as stipulated by the Ministry of Environment, Forest and Climate Change, which emphasize the need to avoid or minimize waste [15]. New environmentally friendly methods such as green synthesis and recycling methods such as chemical precipitation and adsorption are therefore gradually being adopted. Life cycle assessments are now considered as very important tools in the assessment of the sustainability and environmental impacts of AuNP applications [16]. In order to increase compliance and ensure that everyone knows the requirements, training is important and should be done through workshops and training sessions.

However, there is still a lot of variation in the regulations among nations. To address this gap, there is a need for cooperation to develop global standards to prevent, control and minimize the risks associated with AuNP exposure and other nanomaterials, thus requiring the cooperation of international regulatory organizations and scientific associations. In order to achieve the best results in the management of AuNP risks, the various aspects of handling and disposal must be well understood and managed to ensure that safety is coordinated with environmental protection and innovation.

8.2.2 Ethical Implications of Nanomedicine

The contribution of AuNPs in nanomedicine generates numerous new technologies and ethical dilemmas that are worthy of in-depth ethical scrutiny and multidisciplinary discourse. In this context, these nanoparticles have the potential to significantly enhance the specificity of drug delivery and diagnostic systems, and may revolutionize the treatment of patients in the near future. However, ethical questions arise around safety, equity and informed consent. Since there are still many unknowns regarding the long term health effects of AuNPs, there is a critical need to conduct more systematic toxicological studies [19, 20]. The high costs of producing AuNPs also raise equity concerns about the economic accessibility of such advanced medical technologies, which may be unattainable in financially challenged societies.

In this regard, informed consent protocols must be systematically developed to explain all the potential risks of the study to the patients including the possibility of unexpected adverse effects [21]. The use of AuNPs in genetic and molecular treatments also raises another layer of bioethical complexity regarding the impact on human genetics, and there is a pressing demand for more rigorous bioethical analysis. In addition, current regulatory systems need to be enhanced to guarantee that more emphasis is given to issues of transparency, inclusivity, and ethical oversight in the research and clinical implementation of nanomedicine.

For a responsible path forward, ethical reflection and management of AuNPs must be addressed and discussed thoroughly by ethicists, scientists and policymakers [22]. These discussions should lead to the integration of technological development with ethical, moral and humanitarian principles so that the integration of AuNPs in healthcare is not only an improvement in the therapeutic outcomes but also a reinforcement of justice and ethical integrity. Thus, the present work aims to contribute to the discussion of the ethical issues associated with the application of AuNPs in nanomedicine in order to ensure the sustainable development of this field.

8.3 Emerging Trends and Innovations

Recent breakthroughs in the area of AuNP research highlight advancements in greener, more sustainable biosynthetic methods that enable the synthesis of AuNPs via biological systems, such as plant extract and microbes. In parallel, machine learning methods have also improved the selection of nanoparticle sizes, shapes, and other physical parameters for specific target settings. Advances in plasmonic sensing have led to the creation of ultra-sensitive detection platforms. The attachment of AuNPs with a large number of organic and inorganic substances has increased their functional versatility and they are now finding new applications in biomedicine, environmental monitoring, and catalysis. Furthermore, the incorporation of nanostructures through 3D printing has greatly enhanced material properties and newly emerging theranostic AuNPs have shown substantial application potential in various fields.

8.3.1 Use of Hybrid Nanomaterials

Hybrid nanomaterials, comprising AuNPs in association with organic, polymeric, or inorganic matrices have found their application in a wide range of scientific disciplines. These materials harness the unique properties of AuNPs including the surface plasmon resonance to enhance the functionalities of the systems for catalysis, drug delivery and sensing [23, 24]. For example, gold-polymer hybrids have better biocompatibility and enabled controlled drug delivery in nanomedicine. AuNPs incorporated in silica matrices, or linked to metal oxides, enhance catalytic performance and facilitate environmental remediation.

New strategies of hybridization, however, incorporate carbon based materials such as graphene and carbon nanotubes, with AuNPs [25, 26] to enhance conductivity for electronic purposes. Similarly, AuNP—semiconductor nanohybrids are also found to have potential in photocatalysis by improving light harvesting and charge transport. The charge transport ability of these hybrid systems is further investigated for their application in solar energy conversion and environmental pollution mitigation. However, biohybrids, which are AuNPs-based systems along with biomolecules like proteins or DNA [27], are used in biosensing and medical diagnosis. They can determine low-abundance biomarkers with a high level of accuracy. Current trends also include the integration of hybrid nanomaterials into flexible electronics and wearable devices. However, there are still some problems, such as scalability, reproducibility, and longevity. Solving these is critical to improving the availability and usability of hybrid nanomaterials in advanced manufacturing and emerging industries.

8.3.2 Quantum Dots and Multifunctional Nanoparticles

Quantum dots (QDs) and multifunctional nanoparticles (MNPs) are at the top of the nanotechnology field due to their excellent electronic and optical properties. QDs are semiconducting nanoparticles, and based on the size of the particles, there is quantum confinement that leads to selective photoluminescence [28]. This tunability is crucial for the use of QDs in various applications including, bioimaging, and photovoltaics, light emitting diodes, among others. The ability to produce several excitons from a single photon is especially significant for the development of new types of solar cell devices that can operate with higher efficiency.

On the other hand, MNPs, for instance, AuNPs are developed to incorporate diagnostic, therapeutic, and drug delivery functions into a single system [19]. AuNP-based theranostic agents demonstrate this integration by improving the accuracy of cancer diagnosis and the efficacy of the treatment simultaneously. These nanoparticles are often constructed in a sophisticated core–shell structure wherein, the gold core possesses SPR properties, and a shell of polymers or other nanoparticles enables targeted delivery mechanisms.

The integration of QDs and AuNPs into hybrid structures [29, 30] is attractive because of the optical and plasmonic sensitivities that are greatly enhanced for biosensing and spectroscopic analyses. However, there are some critical challenges, including biocompatibility, cytotoxicity, and environmental impact, that are still present and are a major issue for translating these nanoparticles to clinical practice. Future work must prioritize the development of environmentally friendly synthesis and enhanced durability in order to fully realize the transformative capabilities of QDs and MNPs in the area of science and medicine.

8.4 Future Perspectives and Potential Areas for Exploration

Future directions in AuNP research are increasingly aligned with the principles of sustainable development, technological convergence, and multidisciplinary integration thereby defining the next era of nanotechnology. A major research trajectory involves the advancement of green synthesis approaches, which are expected to reduce environmental impact and ecological toxicity. AuNPs are emerging as multifunctional nanocarriers in personalized nanomedicine, enabling the rational design of specific treatment plans tailored to individual patient profiles. Concurrently, developments in hybrid systems and quantum dot integration will enhance the sensitivity, selectivity, and response time of biosensing, environmental monitoring, and energy conversion technologies. Furthermore, the integration of nanoparticles with intelligent devices and flexible electronics paving the way for the development of novel. Thus, solutions to the problems of scalability, reproducibility, and ethical issues will be critical for successful clinical and industrial translation, and thus require collaborative intervention between materials science, biotechnology and computational modelling simulation.

8.4.1 Personalized Nanomedicine

The integration of AuNPs in nanomedicine has significantly transformed the concepts and practices of modern medical science, and it is accompanied by both tremendous potential and considerable ethical issues. Emerging paradigms in personalized nanomedicine, characterized by the development of individualized therapeutic regimens based on a patient's genomic and molecular profile, demonstrate the potential of these advances in improving the accuracy of diagnosis and treatment. AuNPs are employed to improve the delivery of chemotherapeutic agents, including drugs designed to improve cancer cells specifically, thus reducing the impact on other non-target tissues of the body and increasing the therapeutic index [12, 14].

However, there are several ethical dimensions that must be critically examined. Foremost among these is the concern regarding the biosafety of AuNPs. There remains limited understanding of the effects of bioaccumulation and environmental toxicity associated with AuNP exposure. Therefore, it is critically important to carry out systematic toxicological assessments ensure safe translation into clinical practice [19, 20]. Moreover, the cost of AuNP-based therapeutic platforms raises pressing concerns related to health equity, as the economic burden may deny economically vulnerable patients access to important medical innovations.

Furthermore, established informed consent protocols require reconfiguration to ensure that the patient is fully aware of the risks and benefits of AuNP therapies, including those that involve new technologies like CRISPR-Cas9, a gene editing tool. This convergence presents a compelling opportunity for precision medicine, but it necessitates ethical governance to avoid unregulated genomic modifications and potential misuse for enhancement or non-therapeutic purposes [21, 22].

In parallel, the regulatory frameworks must also be updated to incorporate provisions for transparency, accountability, and inclusiveness. It is important to foster collaboration among ethicists, biomedical researchers, and regulatory policymakers to ensure that these innovations are consistent with human values. Thus, it is imperative to foreground the ethical aspects of nanomedicine in order to achieve the full translational potential of AuNPs for the benefit of humanity while upholding the values of equity, safety, and ethical practice in the context of global healthcare systems.

8.4.2 Integration into Smart and Flexible Devices

AuNPs are at the cutting edge of the development of flexible, innovative technologies with user-friendly designs. AuNPs' outstanding plasmonic and conductive capabilities make them suitable for flexible sensors, wearables, and bioelectronics [31, 32]. The key application is in biosensors based on AuNPs, owing to their exceptional sensitivity, rapid signal transduction, and versatility in detecting biomarkers, pollutants, and pathogens.

In flexible electronics, AuNPs contribute substantially to conductivity and stability [32, 33] of various electronic platforms, which is crucial for the development of wearable devices capable of continuous, real-time physiological monitoring. The commercialization of electronic products, such as electronic skin (e-skin), capable of tactile sensing and pressure detection, will revolutionize prosthetic and robotic systems. In addition, AuNP films work by utilize SPR for intelligent optical components to dynamically modulate light pathways and optimize photonic energy harvesting.

Furthermore, the use of AuNP-based inks in 3D printing is instrumental [34] for the production of electric and electronic components with high customizability, reduced material wastage, and lower production costs. The integration of AuNPs into smart textiles and flexible batteries demonstrates their importance in consumer electronics and energy storage devices. However, several challenges persist, including mechanical robustness, cost-effective production, and process integration. To address these challenges, nanomanufacturing, material hybridization, and large-scale synthesis of AuNPs must be further developed to support the growing demand for intelligent, sustainable AuNP-enabled technologies.

References

1. Anik, M.I., Mahmud, N., Al Masud, A., Hasan, M.: Gold nanoparticles (GNPs) in biomedical and clinical applications: a review. Nano Select **3**(4), 792–828 (2022). https://doi.org/10.1002/nano.202100255
2. Ramagiri, K.K., Kar, A.: Environmental impact assessment of alkali-activated mortar with waste precursors and activators. J. Build. Eng. **44**, 103391 (2021). https://doi.org/10.1016/j.jobe.2021.103391
3. Wani, I.A., Jain, S.K., Khan, H., Kalam, A., Ahmad, T.: Gold nanoparticles as efficient catalysts in organic transformations. Curr. Pharm. Biotechnol. **22**(6), 724–732 (2021). https://doi.org/10.2174/1389201022666210218195205
4. Jeevanandam, J., Kiew, S.F., Boakye-Ansah, S., Lau, S.Y., Barhoum, A., Danquah, M.K., Rodrigues, J.: Green approaches for the synthesis of metal and metal oxide nanoparticles using microbial and plant extracts. Nanoscale **14**(7), 2534–2571 (2022). https://doi.org/10.1039/D1NR08144F
5. Salem, S.S., Fouda, A.: Green synthesis of metallic nanoparticles and their prospective biotechnological applications: an overview. Biol. Trace Element Res. (2021).https://doi.org/10.1007/s12011-020-02138-3
6. Bharadwaj, K.K., Rabha, B., Pati, S., Sarkar, T., Choudhury, B.K., Barman, A., Mohd Noor, N.H.: Green synthesis of gold nanoparticles using plant extracts as beneficial prospect for cancer theranostics. Molecules **26**(21), 6389 (2021).https://doi.org/10.3390/molecules26216389
7. Santhosh, P.B., Genova, J., Chamati, H.: Green synthesis of gold nanoparticles: an eco-friendly approach. Chemistry (2022).https://doi.org/10.3390/chemistry4020026
8. Khizar, S., Zine, N., Errachid, A., Jaffrezic-Renault, N., Elaissari, A.: Microfluidic-based nanoparticle synthesis and their potential applications. Electrophoresis **43**(7–8), 819–838 (2022). https://doi.org/10.1002/elps.202100242
9. Saikia, A., Newar, R., Das, S., Singh, A., Deuri, D.J., Baruah, A.: Scopes and challenges of microfluidic technology for nanoparticle synthesis, photocatalysis and sensor applications: a comprehensive review. Chem. Eng. Res. Des. **193**, 516–539 (2023). https://doi.org/10.1016/j.cherd.2023.03.049
10. Yao, F., Zhu, P., Chen, J., Li, S., Sun, B., Li, Y., Chen, Q.: Synthesis of nanoparticles via microfluidic devices and integrated applications. Microchim. Acta **190**(7), 256 (2023). https://doi.org/10.1007/s00604-023-05838-4
11. Karnwal, A., Kumar Sachan, R.S., Devgon, I., Devgon, J., Pant, G., Panchpuri, M., Ahmad, A., Alshammari, M.B., Hossain, K., Kumar, G.: Gold nanoparticles in nanobiotechnology: from synthesis to biosensing applications. ACS Omega **9**(28), 29966–29982 (2024). https://doi.org/10.1021/acsomega.3c10352
12. Niederberger, M., Pinna, N.: Metal Oxide Nanoparticles in Organic Solvents: Synthesis, Formation, Assembly and Application. Springer (2009).https://doi.org/10.1007/978-1-84882-671-7

13. Luque, R., Varma, R.S. (Eds.): Sustainable preparation of metal nanoparticles: Methods and applications. Royal Soc. Chem. (2012). https://doi.org/10.1039/9781849735469
14. Rai, M., Posten, C. (Eds.): Green biosynthesis of nanoparticles: Mechanisms and applications. CABI (2013)
15. Sabela, M., Balme, S., Bechelany, M., Janot, J.-M., Bisetty, K.: A review of gold and silver nanoparticle-based colorimetric sensing assays. Adv. Eng. Mater. **19**(7), 1700270 (2017). https://doi.org/10.1002/adem.201700270
16. Feldheim, D.L., Foss, C.A.: Metal nanoparticles: Synthesis, characterization, and applications. Marcel Dekker (2001). https://doi.org/10.1201/9780367800475
17. Gomez-Villalba, L.S., Salcines, C., Fort, R.: Application of inorganic nanomaterials in cultural heritage conservation, risk of toxicity, and preventive measures. Nanomaterials (2023). https://doi.org/10.3390/nano13091454
18. Wahab, A., Muhammad, M., Ullah, S., Abdi, G., Shah, G.M., Zaman, W., Ayaz, A.: Agriculture and environmental management through nanotechnology: eco-friendly nanomaterial synthesis for soil-plant systems, food safety, and sustainability. Sci. Total Environ., 171862 (2024).https://doi.org/10.1016/j.scitotenv.2024.171862
19. Al-Thani, A.N., Jan, A.G., Abbas, M., Geetha, M., Sadasivuni, K.K.: Nanoparticles in cancer theragnostic and drug delivery: a comprehensive review. Life Sci., 122899 (2024). https://doi.org/10.1016/j.lfs.2024.122899
20. Hossain, A., Rayhan, M.T., Mobarak, M.H., Rimon, M.I.H., Hossain, N., Islam, S., Al Kafi, S.A.: Advances and significances of gold nanoparticles in cancer treatment: a comprehensive review. Results in Chem., 101559 (2024).https://doi.org/10.1016/j.rechem.2024.101559
21. Ehidiamen, A.J., Oladapo, O.O.: Enhancing ethical standards in clinical trials: A deep dive into regulatory compliance, informed consent, and participant rights protection frameworks. World J. Biol. Pharmacy Health Sci. **20**(01), 309–320 (2024). https://doi.org/10.30574/wjbphs.2024.20.1.0788
22. Schroeder, S.A.: An ethical framework for presenting scientific results to policy-makers. Kennedy Inst. Ethics J. **32**(1), 33–67 (2022). https://doi.org/10.1353/ken.2022.0002
23. Hang, Y., Wang, A., Wu, N.: Plasmonic silver and gold nanoparticles: shape-and structure-modulated plasmonic functionality for point-of-caring sensing, bio-imaging and medical therapy. Chem. Soc. Rev. (2024).https://doi.org/10.1039/D3CS00793F
24. Kumalasari, M.R., Alfanaar, R., Andreani, A.S.: Gold nanoparticles (AuNPs): A versatile material for biosensor application. Talanta Open (2024).https://doi.org/10.1016/j.talo.2024.100327
25. Fritea, L., Banica, F., Costea, T.O., Moldovan, L., Dobjanschi, L., Muresan, M., Cavalu, S.: Metal nanoparticles and carbon-based nanomaterials for improved performances of electrochemical (Bio) sensors with biomedical applications. Materials **14**(21), 6319 (2021). https://doi.org/10.3390/ma14216319
26. Lopes, R.C., Rocha, B.G., Maçôas, E.M., Marques, E.F., Martinho, J.M.: Combining metal nanoclusters and carbon nanomaterials: opportunities and challenges in advanced nanohybrids. Adv. Coll. Interface. Sci. **304**, 102667 (2022). https://doi.org/10.1016/j.cis.2022.102667
27. Lee, J.W., Choi, S.R., Heo, J.H.: Simultaneous stabilization and functionalization of gold nanoparticles via biomolecule conjugation: Progress and perspectives. ACS Appl. Mater. Interfaces (2021).https://doi.org/10.1021/acsami.1c10436
28. Das, N., Panda, S., Das, D.K., Nayak, S.K., Parthasarathy, S., Patnaik, P., Behera, S.: Unraveling quantum dots: Trailblazers in nanoscience and beyond. In: Discovery, Disruption, and Future Implications of Nanomaterials, pp. 183–218. IGI Global (2024)
29. Raj, D., Kumar, A., Kumar, D., Kant, K., Mathur, A.: Gold–graphene quantum dot hybrid nanoparticle for smart diagnostics of prostate cancer. Biosensors (2024).https://doi.org/10.3390/bios14110534
30. Seaberg, J., Montazerian, H., Hossen, M.N., Bhattacharya, R., Khademhosseini, A., Mukherjee, P.: Hybrid nanosystems for biomedical applications. ACS Nano **15**(2), 2099–2142 (2021). https://doi.org/10.1021/acsnano.0c09382

31. Chang, L., Linbo, H., Joanna, W., Jiacheng, W., Guosheng, S., Jianghong, R.: Engineering of magnetic nanoparticles as magnetic particle imaging tracers. Chemical Society Reviews. **50**(14) 8102–8146 (2021). https://doi.org/10.1039/D0CS00260G
32. Mohamed, M., Ghobashy Sh. A, Alkhursani Ha. A., Alqahtani, Ta. K., El-damhougy, Mo. M.: Gold nanoparticles in microelectronics advancements and biomedical applications. Materials Science and Engineering: B. **301** 117191 (2024). https://doi.org/10.1016/j.mseb.2024.117191
33. Jisun, I., Gustavo F., Trindade T., Thuy, Q., Ali, S., Feiran, W., Jonathan, A., Lyudmila, T., Clive J., R., Ricky, W., Richard, H., Christopher, T.: (2022) Functionalized Gold Nanoparticles with a Cohesion Enhancer for Robust Flexible Electrodes ACS Applied. Nano Materials. **5**(5) 6708–6716 (2022). https://doi.org/10.1021/acsanm.2c00742
34. Pengxuan, Z., Youbin, D., Guangya, X., Yani, L.: Nanoparticle-Assisted Sonosensitizers and Their Biomedical Applications. International Journal of Nanomedicine Volume. **16** 4615–4630 (2021). https://doi.org/10.2147/IJN.S307885

MIX
Papier aus verantwortungsvollen Quellen
Paper from responsible sources
FSC® C105338

If you have any concerns about our products,
you can contact us on
ProductSafety@springernature.com

In case Publisher is established outside the EU,
the EU authorized representative is:
Springer Nature Customer Service Center GmbH
Europaplatz 3, 69115 Heidelberg, Germany

Printed by Libri Plureos GmbH
in Hamburg, Germany